Infinity's Rainbow

Infinity's Rainbow

The Politics of Energy, Climate and Globalization

Michael P. Byron

Algora Publishing
New York

ISBN-13: 978-0-87586-509-6 (trade paper)
ISBN-13: 978-0-87586-510-2 (hard cover)
ISBN-13: 978-0-87586-511-9 (ebook)

Library of Congress Cataloging-in-Publication Data —

Byron, Michael P., 1956-
    Infinity's rainbow: the politics of energy, climate, and globalization / Michael
P. Byron.
        p. cm.
    Includes bibliographical references and index.
    ISBN-13: 978-0-87586-509-6 (trade paper : alk. paper)
    ISBN-13: 978-0-87586-510-2 (hard cover : alk. paper)
    ISBN-13: 978-0-87586-511-9 (ebook)
    1. Political ecology—United States. 2. Energy policy—Political aspects—
United States. 3. Energy policy—Economic aspects—United States. 4. Sustainable
development—Political aspects—United States. I. Title.

JA75.8.B85 2007
333.790973—dc22
                            2006029669

To my wife and partner Ramona, without whose support, encouragement, ideals and ideas, and contributions this work would not have been possible.

Also dedicated to all those who will "Move the Earth" in the years to come.

Acknowledgements

I would like to thank the following people for their support and assistance during the writing of this book:

My wife Ramona Byron for her many suggestions and careful proofreading of my manuscript.

Maegan Prentiss, Charles Stidd, Barbara Parcells, Wayne Strache, Carolyn Baker, Lori Price, Felix Geyer, Sorab Ghandi, Richard Keeley, Anna O'Cain, Michael Bressler, and Mark Hull-Richter, for agreeing to read my manuscript, or otherwise contributing to it.

Gregory Benford, for his insights and encouragement — we may not always agree, however I deeply appreciate your insights.

I would also like to thank Martin DeMers at Algora Publishing for agreeing to publish my book.

TABLE OF CONTENTS

Table of Contents

PROLOGUE

It has often been said that, if the human species fails to make a go of it here on the earth, some other species will take over the running. In the sense of developing intelligence this is not correct. We have or soon will have, exhausted the necessary physical prerequisites so far as this planet is concerned. With coal gone, oil gone, high-grade metallic ores gone, no species however competent can make the long climb from primitive conditions to high-level technology. This is a one-shot affair. If we fail, this planetary system fails so far as intelligence is concerned. The same will be true of other planetary systems. On each of them there will be one chance and one chance only.
— Fred Hoyle[1]

We live at the decisive moment in all of human history — decisive not only for one culture or another, not only for the "developed world," but for all of humanity.

In a vivid metaphoric sense, civilization is now in a condition analogous to that of the astronauts on the space shuttle Columbia as it reentered Earth's atmosphere on the morning of February 1, 2003. An eerie videotape of those last moments was found amid Columbia's debris afterwards. Onboard Columbia, the lights are on, the air is circulating, and all seems well.

The four astronauts, seen on the flight deck of shuttle, marvel together at the sight of the white-hot plasma flowing outside around them. They are unaware that this plasma is patiently eating away at the damaged left wing of their spaceship. There is only one hint of the slowly unfolding catastrophe: the ship's guidance thruster begins firing ever more frequently and thunderously, as the computers that are actually flying the vehicle sense the asymmetric drag caused by the eroding left wing and vainly try to compensate for it.

---

1. Hoyle, Fred, *Of Men and Galaxies.* University of Washington Press, Seattle, WA, 1964, pp. 73.

The sights and sounds of these repeated thruster firings are clearly evident in the videotape. The laws of physics, those immutable and cold equations of nature, have decreed that in just moments all seven astronauts aboard the Columbia will die catastrophically. No sensor sounds the alarm because the sensors had already burned up. Although the disaster is unfolding in slow-motion around them, the astronauts have no direct way to detect it. All still appears reasonably normal as the tape abruptly ends.[2]

Civilization at this very moment is in nearly the same situation. The world-spanning industrial civilization now seems doomed to certain catastrophe. With this book I hope to provide you, the readers, with an understanding of the problems and with strategies to for the future, and a look toward the renewal of civilization itself.

The linked crises which are bearing down upon humanity are not caused by external disturbances. There are no cruel gods who have determined to torment us for their amusement. Rather, these crises are self-caused and originate within civilization. They have sprung from our deepest values, beliefs, and unquestioning assumptions about reality itself. As Cassius observes in the Shakespeare play Julius Caesar: "The fault dear Brutus is not in our stars but in ourselves."[3]

Anatomically modern humans have existed upon the earth for perhaps 200,000 years. Civilization has only emerged in the past ten millennia; it is characterized by dense, settled populations centered in cities that do not produce their own food, and with people differentiated by occupation and social class.

Civilization used to depend upon human and animal muscle power. Only the industrialization of the past 200 years has substituted muscle power with concentrated sources of energy, primarily hydrocarbon fuel, energy in the forms of coal, oil, and natural gas.

As industrialization based primarily upon these fossil fuels has spread across the planet, a globalized economy has emerged. We seem to be standing at the very summit of human achievement with power over nature, and with wealth and opportunity for all. A closer look reveals that we have arrived not only at the summit, but also at the edge of a precipice — a yawning chasm in

---

2. Washington Post.com, Feb. 28, 2003, Recovered Video Fragment Shows Crew During Reentry, http://www.washingtonpost.com/wp-srv/mmedia/nation/022803-4v.htm, NASA has the video online at: http://realserver1.jpl.nasa.gov:8080/ramgen/sts107.rm and also at: http://vstream1.ksc.nasa.gov/ramgen/odv/ksc_direct/sts107/jsc_022803_crew.rm and finally at: http://science.ksc.nasa.gov/cgi-bin/rrg2.pl?video/shuttle/missions/sts-107/sts107a.rm Note: The actual video can be watched online. A reconstruction of Columbia's final seven minutes which synchronizes ground video with Mission Control dialogue is available at: http://www.chrisvalentines.com/sts107/realtime_play.html

3. Enotes.com, Julius Caesar, http://www.allshakespeare.com/jc/270

human history. Civilization's foundation is fatally insecure and its collapse is imminent.

All of civilization is predicated upon one mostly unspoken assumption: that limitless supplies of cheap hydrocarbon energy will always be available.

Corollaries to this core assumption include the assumptions that, if hydrocarbon energy ever does become scarce, markets will instigate the development of substitute sources of energy; and that science and technology will be able to rapidly develop these substitute sources of energy. A third assumption is that human actions have little or no effect upon the weather, and on Earth's ability to maintain the conditions necessary for human life to flourish. Finally, it is presumed that the political leadership will respond quickly and adequately to problems which affect mankind's very existence.

Unfortunately, *ALL of the above are false.* The global hydrocarbon reserves which we are so recklessly squandering took several hundred million years to accumulate. Once they're gone, they are gone forever. The current high-energy industrial civilization can only occur once in the lifetime of the planet. What comes next is anyone's guess, but the adjustment period, at least, is likely to be disastrous and if no adequate adjustment can be made, then the initial disaster must lead to a dismal end for mankind.

Hydrocarbon energy powers industries, automobiles, and aircraft. It heats homes. It makes possible industrial-scale agriculture. Indeed, fertilizers are made from natural gas, and pesticides from petroleum. So hydrocarbon-intensive is modern agriculture that for every one calorie of food produced, about ten calories of irreplaceable hydrocarbon energy is expended. We are eating hydrocarbon energy!

Simultaneously, the heat-trapping greenhouse gases are disrupting the planet's thin and finite atmosphere, leading to the ever more rapid warming of the planet and destabilizing the weather patterns.

At this very moment, we are about to begin to run out of these irreplaceable hydrocarbon energy sources. Estimates indicate that we are just about reaching the midpoint of world petroleum production — the point at which one half of all the oil that can ever be produced will have been produced. In the near future there will be ever less oil, less gasoline, less kerosene, less jet fuel, etc., produced each year than was produced the previous year.

This peaking of oil production — commonly referred to as "peak oil" — is occurring at a time when demand for hydrocarbon energy is increasing at a rate of over two percent per year, compounded, as large and populous nations such as China and India rapidly industrialize. In fact, the *rate* of increase in demand is increasing rapidly. Obviously, this has dire political consequences for the peace and stability of the planet.

Natural gas production in North America is also about to peak. Globally, the natural gas peak is only about a decade away. In any event, even if there was gas to buy, it would take at least a decade to build the immense tankers, specialized ports and refineries required to import natural gas to North America. The option of importing natural gas in order to stave off impending oil depletion is impractical.

It is true that coal exists in large quantities around the globe. Indeed, the United States possesses the planet's largest reserves. However, coal cannot substitute for all uses of petroleum, and would itself be totally depleted by several decades of intensive usage. More significantly, it is by far the "dirtiest" of the hydrocarbon fuels. Its widespread use would kick global warming further into overdrive.

Also, the political system in the United States has been decisively captured by multinational corporations which profit immensely from keeping the global political economy based on hydrocarbon energy. The influence of these corporations upon the world's governments cannot be overstated; and it is an influence that operates to maximize profits, not to maximize what is in the best interests of humanity. As a consequence, government itself has become a major part of the problem to addressing these mounting and imminent crises. It is not part of the solution.

And so here in the early 21st century we find ourselves standing at a turning point in human history. The choices we make will irrevocably determine the fates of all future humans living in all future ages. There is no second chance!

> The end of the Roman West witnessed horrors and dislocations of a kind I sincerely hope never to have to live through; and it destroyed a complex civilization, throwing the inhabitants of the west back to a standard of living typical of prehistoric times. *Romans before the fall were as certain as we are today that their world would continue for ever substantially unchanged. They were wrong. We would be wise not to repeat their complacency.*[4]

OVERVIEW

Crisis is coming. However, it is very important to understand that it is not something that is being done to us; rather it is something that we are doing to ourselves. We, Homo Sapiens Sapiens, are the culprit, and not some invisible, cruel gods.

To understand the causes and consequences of our actions, we must first understand ourselves. We must understand how we create our realities, and how this affects the ways that we respond to crisis. Additionally, we must understand systems theory because what we do affects everything around us.

---

4. Ward-Perkins, Brian, *The Fall of Rome and the End of Civilization*, Oxford University Press, New York, NY, 2005, pp. 182

This book is intended to give you, the reader, insight into how and why these crises are bearing down upon us, and what their effects will be. It is further intended to empower you to participate in the creation of a new approach that will support sustainable practices and provide a decent quality of life without destroying the world's resources.

Early 21st century civilization is a human-created, human-centered, world-spanning, complex adaptive system, containing within it nested political, social, and economic systems organized at multiple levels from the individual to the world system.

This civilization is itself nested within the earth's four-billion years old biological system — the biosphere. This gradually-evolved and finely-tuned system has made the planet habitable for eons. Perhaps unsurprisingly, we take it for granted and assume that it will always be there to provide for us.

The biosphere is in turn nested within the natural, non-living, physical systems of the earth itself. These natural systems include the planet's plate tectonic and volcanic cycles and its hydrological cycles. These in turn — especially the earth's hydrological cycle — are heavily dependent upon the amount of sunlight reaching the planet. Ultimately, the earth, sun, and solar system are nested within the subtle energy fields of the entire surrounding universe.

My focus is on the crises caused or exacerbated by humans, and that are bearing down upon civilization. For this book's purposes, humans are at the very center of this concentric set of nested systems. This central position of humans means that understanding the manner in which our minds, or more precisely, our brains, process, store and organize information into ordered patterns is crucially important. This is because these structured patterns of information determine the very nature of large-scale human social and societal organization — our beliefs, dogmas and cultures — from the individual level all of the way up to the level of our global civilization itself.

These belief structures in turn determine what we do — and what we don't do. Furthermore, the nature of these cognitive building blocks determines what types of belief structures can be built in the future. As we shall see, the methods by which we create our ordered patterns of understanding the world around us determine how we respond to fundamental crises, both as individuals and as civilizations.

This book is divided into three sections.

Part I, "Conceptual Foundations," explains two vital topics: how our brains process and organize information, and systems theory. It is a bit more technical than the remaining several chapters of the book.

Part II, "The Crises," investigates the linked crises which threaten the collapse of civilization. Our high-energy, world-girdling civilization runs mainly on hydrocarbon energy sources: Coal, oil, and natural gas. These irreplaceable energy sources are being consumed at an accelerating rate. They are about to go into permanent and irreversible decline.

Part III, "Survival & Renewal," addresses the possible impending catastrophe and ponders the essential qualities of a better civilization we might wish to build.

Humanity stands facing a historical chasm. A great discontinuity is about to separate the future from the past. Ahead lie decades and perhaps centuries of turmoil and tumult.

Those who control the global corporations, with their eyes only on the bottom line, bear the ultimate responsibility for the impending collapse of civilization. Even now, citizens are increasingly reduced to politically-disempowered consumers. The now near-total control over the consolidated and corporatized media gives them control over the public's understanding of reality, and ensures that most citizens are sleepwalking towards doom.

By understanding the crises that are by now perhaps unavoidable, and by preparing ourselves, we can hope to minimize the suffering of the adjustment when the irresponsible practices of the present become absolutely unsustainable and force us to change our ways, radically reducing consumption of resources we still treat as infinite even today.

With prudent planning and action, by acting methodically and with purpose in coordination with many others, we can plant the seeds of a future that will offer a much greater quality of life for our descendents than that offered by today's soulless, materialistic, global corporate oligarchy.

Viewed from this context, the opportunity to correct the errors of the present age and of our social and political order, and to bequeath this new order to future ages, is an exciting challenge. It is time we got things right and designed a more mature civilization.

To begin to develop a picture of how and why these interrelated crises have been allowed to develop, and to understand what we can do, we next need to take a careful look at three related concepts as they apply to human beings: cognition, crisis and systems theory. All deal intimately with the very nature of reality as it operates for humans organized into large scale entities such as nations, economies, and ultimately, globalized civilization taken as a whole. Then we shall investigate the crises bearing down upon us, and what we can do to protect ourselves, our families and our communities. Finally we will investigate how our actions can bring about a better tomorrow for our descendents.

Beyond our personal survival and the survival of those closest to us, we can begin to plan for the rebirth of civilization — of a better, far more humane and fulfilling, civilization than the tragically flawed one around us which is now racing towards its utter doom in the years just ahead. This topic shall form the final portion of the book. We can act to create the seeds, the nodes, of a democratic and humane civilization, based upon a stable foundation of renewable energy sources, sustainable agriculture, smaller, closer, more caring communities, which is predicated upon a social foundation of individual and human rights.

Ultimately, by changing the way that we think about our external reality, we can change how we interact with one another and with nature. Such a change may yet give a more humane human civilization a new lease on life and eventually still lead us to the stars in ages to come. Surely such a world is not to be feared but rather, it is to be embraced.

The alternatives are nightmarish! We must make the effort!

PART I
CONCEPTUAL FOUNDATIONS

CHAPTER 1. CRISIS AND COGNITION

We don't see things as they are, we see them as we are.
— Anaïs Nin [5]

CRISIS DEFINED

Why do individuals and their aggregates — groups, religions, societies, civilizations change over time? Most people probably assume that historical change is usually gradual; and indeed measured, incremental change certainly does occur. Yet seen from the long view of human history (and pre-history), fundamental change usually occurs suddenly and discontinuously.

When, for any reason, old, established techniques for dealing with events suddenly cease to produce the expected, beneficial results which they have reliably produced in the past, our future well-being, and perhaps even our very existence are acutely threatened. The old ways no longer work, while the present reality is intolerable. And since our former coping strategies have failed us, we are forced to innovate. I define such circumstances as comprising a "crisis."

I am using this term "crisis" in a restricted, technical sense which may differ from its wider and more general everyday usage. Unless otherwise indicated this is the sense in which I shall employ the word "crisis" throughout this book.

I use *crisis* to mean the occurrence of a set of circumstances which cannot be dealt with by using existing behaviors and strategies, but which must be dealt with quickly in order to avert serious harm or death from occurring. Crisis *causes* behavioral adaptation.

---

5. Wisdom Quotes: Anaïs Nin, http://www.wisdomquotes.com/000128.html

Crisis compels individuals and systems composed of individuals, such as nation-states, civilizations, etc, to become more open to new ideas and ways of doing things. In fact, we find that throughout human history, fundamental change has occurred not gradually and incrementally, but rather as large sweeping political, social, economic, technological and religious reconfigurations, as a direct result of crisis.

Looked at in this way, human history can be understood as having developed in response to a series of existential crises. The first of these great crises occurred as a result of a rapid climate change during the ending of the last Ice Age, about 10,000 years ago.

Our species Homo Sapiens Sapiens emerged in Africa by about 200,000 years ago. By about 90,000 years ago there is evidence for these early humans having reached the Middle East. By perhaps 40,000 years ago they had spread across the habitable regions of Eurasia. About 12,000 years ago the world began to experience rapid and dramatic climatic warming as the most recent Ice Age came to an end about 10,000 years ago. Our planetary climate has been fairly stable ever since.

All of the elements needed to create civilization were now in play: A large-brained species; sufficiently dense human populations for economic specialization and culture to develop; and complex language to allow for ideas to be communicated from one individual to another and from one generation to another, which made complex cultures possible. However, reliance upon a hunter-gatherer lifestyle mandated that populations be small and spread out. To make the leap in organization from nomadic hunter-gatherer to civilized, a revolution was required in how available sources of energy were utilized to produce food to sustain larger, settled populations.

Sudden climate change applied selection pressure upon these mutually interacting developments, leading small, tribally organized hunter-gatherer bands to increase in population in those areas which they already occupied, to expand into new, recently habitable areas, and to increase dramatically in population size. These larger populations could no longer be sustained by traditional means of hunting and gathering. Continuing to utilize the old behavioral responses led to famine and to death. Some new behavior was required if life was to be sustained. It was under this selection pressure that agriculture, the foundation of all human civilizations to come, was first adopted.

As one researcher describes this:

> Around 10,000 years ago warming climatic conditions, occurring coincident with significant environmental stress, led to a substantial population increase in the near-east. Populations became too large to support through the old method of hunter-gathering. Just becoming a better hunter-gatherer could not resolve the problem. To survive, a new survival strategy had to be created. This crisis led to the development, and widespread adoption of agriculture.[6]

And so agriculture, a revolutionary innovation, solved the crises of survival brought about by the ending of the last Ice Age. Civilization, a mode of human organization based upon dense populations, specialization of labor, and social and economic stratification, emerged as a direct consequence of agriculture's energy-concentrating innovations.

Civilization requires not only the reliable mass production of food which is made possible by agriculture, it also requires that its members have the ability to think, learn and believe in coordinated ways. Civilization requires a common culture which allows for large numbers of unrelated individuals to live in close proximity to one another in relative peace and order. Culture, technology, and civilization itself are products of our large, complex brains.

In furtherance of our quest to understand how we have come to face the present crises of the 21st century, and to understand what we must do to solve these crises, we need to understand something about how our minds are organized, as it is our minds which have created today's reality.

The seeds of today's mega-crises are ultimately rooted in the soil of our very minds. To understand how this looming disaster has come about and what we can do about it we must first look not to our stars — not to any cruel external arbiter of destiny — but rather we must look deeply into ourselves; for we and we alone, are the ultimate cause of our coming misfortunes...

PARAMETERS OF COGNITION — HOW WE THINK

How do our minds work? What are the relevant parameters of human thinking? This discussion will get just a bit technical. I ask the reader to please hang in there, as it does contain important information directly relevant to our investigation. There is considerable evidence suggesting that knowledge is stored hierarchically and schematically in networks of association by our brains. This means that, for example, smelling salty air while walking on a beach brings to mind earlier memories involving the sea in some manner. Similarly, the smell of a rose's fragrance brings with it memories of walking in flowery meadows, of bouquets of roses, and of long ago romances, of times and places, people and events that we'd long forgotten. These memories if contemplated will bring forth still other tangentially associated memories. Our brains store information hierarchically and schematically in networks of association because:

- It is an efficient and parsimonious usage of limited cognitive resources.
- It has survival value.

6. Wenke, Robert J., Patterns in Prehistory: Humankind's First Three Million Years, Third Edition, Oxford University Press, New York, NY, 1990, pp. 269.

If, for example, while we are out hiking in the woods, we stumble upon a large, growling bear, which suddenly lunges towards us, we simply do not have the time to methodically review our lifetime's accumulation of memories, searching randomly for those that give us insights into what to do. So we don't do that. What we do instead is to recall only those memories that we have that are associated with charging bears: Just what was it that Marlin Perkins, the host of the 1960s TV show "Wild Kingdom," said to do when a bear charges at you? What did the "Crocodile Hunter" say to do on his TV show? Accessing these memories has survival value for us. And so that's what we do (hopefully). Perhaps we recall that most charges by bears are bluffs. Perkins advised us (if I remember correctly) to back away slowly, facing the bear, while avoiding eye contact...

Amazingly, out of a lifetime of memories we can extract just those we need, just when we need them. That is the power of organizing our thinking associationally!

We process and store information in this manner because we have a limited number of cognitive cells called neurons (10 to 100 billion), linked together to form circuits of information processing junctures called synapses. Each neuron possesses up to several thousand of these connections with other neurons, via each neuron's information conducting axons and dendrites.

The human nervous system is composed of about one hundred billion neurons. Each neuron connects with, on average, about one thousand other neurons, forming about ten trillion synapses. The ultimate storage capacity of the human brain has been calculated to be approximately 10^{18} bits of information (1,000,000,000,000,000,000), or one pentillion bits.[7]

Each neuron has a firing speed of about one millisecond. That is, it is able to perform about one thousand operations per second under ideal conditions. Allowing for propagation delays between neurons, this works out to a firing rate of about one hundred times a second. As the human brain is wired into a massively parallel configuration, many neurons are firing simultaneously. To survive in a complex and often dangerous world, people need to be able to process information rapidly. Processes, such as recognition, require roughly 100 to 1,000 synaptic firings. As a general rule:

> This means that processes that take on the order of a second, or less, can involve only a hundred, or so, time steps. Because most of the processes we have studied — perception, memory, retrieval, speech processing, sentence comprehension, and the like, take about a second, or so, it makes sense to impose [what is called] 'the 100-step program constraint.' That is, we seek explanations for these mental phenomena that do not require more than about a hundred elementary sequential operations.[8]

---

7. Kurzweil, Ray, *The Singularity is Near*, Viking Penguin, New York, NY, 2004, pp 127.

Given these limitations, information is generally believed to be stored schematically and associatively. This is so because schematic encoding represents a parsimonious system of information storage and retrieval. In one hundred, or fewer, sequential search operations, any encoded (memorized) information may be retrieved. This allows for a reaction time of one second, or less, and hence, is conducive to survival.

A cognitive schema may be defined as being:

> A spatially, and/or temporally organized structure in which the parts are connected on the basis of contiguities that have been experienced in space and time. A schema is formed on the basis of past experience with objects, scenes, or events, and consists of a set of (usually unconscious) expectations about what things look like, and/or the order in which they occur. The parts, or units, of a schema consist of a set of variables, or slots, which can be filled, or instantiated, in any given instance by values that have greater or lesser, degrees of probability of occurrence attached to them. Schemata vary greatly in their degrees of generality — the more general the schemas, the less specified or the less predictable, are the values that may satisfy them.[9]

According to two social scientific researchers:

> Schemas are not necessarily isolated cognitive structures. Rather, they may be linked to one another through a rich network of hierarchical relationships in which individual schema are 'embedded' in one another, so that the higher-order, more abstract, schemas are characterized in terms of their more concrete, lower-order constituents.[10]

Humans exhibit a fundamental constraint regarding the volume of information they can focus conscious attention upon. One researcher states that:

> The fundamental constraint that underlies all the operations of attention, imposing their essentially selective character, is the limited information processing capability of the brain...Hence according to the predominant view, the basic function, or purpose, of attentional mechanisms is to protect the brain's limited capacity system (or systems) from informational overload.[11]

In other words, conscious awareness may be focused upon some small, albeit situationally relevant, portion of schematically encoded long-term memory.

8. Rumelhart, David E., The Architecture of the Mind: a Connectionist Approach; in *Foundations of Cognitive Science*, Posner, Michael I., Ed., MIT Press, Cambridge, MA, 1991, pp. 135.

9. Schater, Daniel, Memory; in *Foundations of Cognitive Science*, Posner, Michael I., Ed., MIT Press, Cambridge, MA, 1991, pp. 692.

10. Connover, P., & Feldman, S., How People Organize the Political World: A Schematic Model, *American Journal of Political Science*, Vol. 81, 1994, pps. 95-126.

11. Allport, Allan, Visual Attention; in *Foundations of Cognitive Science*, Posner, Michael I., Ed., MIT Press, Cambridge, MA, 1991, pps. 632-33.

Given this model, one social scientist offers several insightful observations. He notes that our senses are highly parallel, possessing, in effect, a large bandwidth, for the input of sensory information regarding "what's out there." The bottleneck for human actors is analogous to that found in a typical computer: The CPU (central processing unit) is only able to process information serially:

> [Information]...must proceed through the bottleneck of attention — a serial, not parallel, process where information capacity is exceedingly small. Psychologists usually call this bottleneck short-term memory and measurements show reliably that it can only hold about six chunks (that is to say, six familiar bits) of information.[12]

Daniel Dennett is the Director of the Center for Cognitive Studies at Tufts University in Massachusetts.[13] His ground breaking research into the workings of human consciousness is in full agreement with this finding. Indeed, he extends it considerably by noting that most information is filtered out before it can ever, even potentially, reach conscious awareness.

Dennett describes the mind as being a sort of virtual computer animated by what he calls "memes." Memes can be succinctly defined as self-perpetuating ideas and beliefs, effectively the mental equivalent of a gene.[14] Viewed in this context, they can be considered as comprising comprehensive, or global, schemas, for any given subject. As such, memes comprise the basic informational unit of cultural transmission.

The concept of memes was originated by the British evolutionary biologist/sociobiologist Richard Dawkins:

> [Memes are] ...a unit of cultural transmission, or a unit of imitation....Examples of memes are tunes, ideas, catch-phrases, clothes, fashions, ways of making pots or of building arches. Just as genes propagate themselves in the gene pool by leaping from body to body via sperm or eggs, so memes propagate themselves in the meme pool by leaping from brain to brain via a process which, in the broad sense can be called imitation. If a scientist heard about a good idea, he passes it on to his colleagues and students. He mentions it in his articles and his lectures. If the idea catches on, it can be said to propagate itself, spreading from brain to brain.[15]

Dennett states:

> Human consciousness is itself a huge complex of memes (or more exactly meme-effects in brains) that can best be understood as the operation of a Von-Neu-

---

12. Simon, Herbert A., Human Nature in Politics; The Dialogue of Psychology With Political Science, *American Political Science Review*, Vol. 82, 1985, pp. 302.

13. The website for the Center for Cognitive Studies contains a wealth of fascinating and informative materials. It is located at: http://ase.tufts.edu/cogstud/index.asp.

14. *Britannica Book of the Year, 1997*, 2006, Dawkins, Richard, Encyclopædia Britannica Premium Service, http://www.britannica.com/eb/article-9112956.

15. Dawkins, Richard, *The Selfish Gene*, Oxford University Press, New York, NY, 1976, pp. 206.

manesque virtual machine implemented in the parallel architecture of a brain that was not designed for any such activities. The powers of this virtual machine vastly enhance the underlying powers of the organic hardware on which it runs, but at the same time many of its most curious features, and especially its limitations, can be explained as by-products of the "kludges" that make possible this curious but effective reuse of an existing organ for novel purposes.[16]

Further developing this theme, in accord with modern empirical findings from brain research, Dennett continues:

Thousands of memes, mostly borne by language, but also by wordless "images" and other data structures, take up residence in an individual brain, shaping its tendencies and thereby turning it into a mind.[17]

Recently Dennett has articulated the concept of what he calls "memetic engineering." He describes this as comprising:

the attempt to design and spread whole systems of human culture, ethical theories, political ideologies, systems of justice and government, a cornucopia of competing designs for living in social groups.[18]

Here we finally see memes as being not just something that passively determines how we see the world around us and interact with it and with other humans, but rather as something that we can actively select. We can choose how we understand the world around us and how we organize ourselves into complex systems with other humans. These choices can be based upon our experiences and learning. We are free to create our destinies. Memes are capable of undergoing evolution according to natural selection. For any environment some are more "fit" than are others. All ideas and beliefs are not equally valid with respect to fitness. Meme complexes can themselves be considered to be complex adaptive systems — a topic about which I shall explain fully and have much more to say about in the next chapter.

SUMMARY OF PARAMETERS OF COGNITION

There are about 10^{11} cognitive elements, known as neurons in the brain. Each neuron is connected to roughly 10^3 others forming about 10^{14} synapses. The brain's ultimate storage capacity (memory capacity) may be as great as 10^{18} bits. Each neuron requires about one millisecond to "fire." Allowing for transmission delays each neuron can fire about 100 times per second. Many neurons can fire simultaneously as the brain processes information in parallel. However, con-

---

16. Dennett, Daniel, *Consciousness Explained*, Little Brown & Company, Boston, MA, 1991, pp. 210.
17. Ibid., pp. 254.
18. Dennett, Daniel, *Freedom Evolves*, Viking Penguin, New York, NY, 2003, pp. 266.

scious processing of information is done serially. Conscious awareness has an information content of about 6-8 "chunks."

Given these information processing constraints, storing memory associatively in schematic arrangements allows for rapid associational retrieval of stored information — a prerequisite for survival in nature. The human brain itself consists primarily of evolutionarily derived, hard-wired, information-processing modules. Conversely the human mind is, in effect, a virtual program, whose fundamental elements are memes which run on the hardware provided by the brain in a manner analogous to how a program runs on the hardware of a computer. In other words, brains are the hardware while memes are the software.

Despite its limitations, this "software" of the human mind is extremely flexible and adaptable, although ultimately bounded by the limitations of the brain upon which it runs. This memetic software can be envisaged as grouping, or organizing, the schematically encoded information which the brain stores.

The serial nature of human conscious awareness, meaning that we can only pay attention to one thing at a time, imposes a fundamental constraint upon the ability of humans to deal with contingencies. Concentrating upon any given task, or information set, must necessarily be done at the expense of other tasks.

People possess substantial memory capacity. This memory information is linked together associatively into schemas. Aggregate schemas for any given subject comprise memes. This form of information storage allows for rapid information retrieval, and utilization, within the 100-step constraints imposed by the brain's biochemistry.

Figure 1 provides a simplified representation of memes, meta-memes, and schemas.

The ability to store information and subsequently to correctly retrieve, interpret, and utilize it, represents learning. As environmental conditions are dynamic, learning must be adaptive. That is, it must be able to respond flexibly, and innovatively, to varying environmental conditions in order to maximize survival. People are capable of altering their beliefs based upon experience. Humans are capable of learning, and given capricious social and natural environments, of adaptive learning.

CRISIS AS SCHEMATIC REPLACEMENT

Given these cognitive parameters, it is reasonable to assume that groups of people develop a comprehensive schema that is shared by all members of the population, and which determines how they collectively respond to environmental challenges. The usual name for this shared meta-schema is "culture." As long as the environment does not change much, learning-based behavioral changes are incremental, if they occur at all. With practice, you get better at doing something, for example. When, and if the environment changes substantially, the previous schematically encoded behavioral learning is no longer useful.

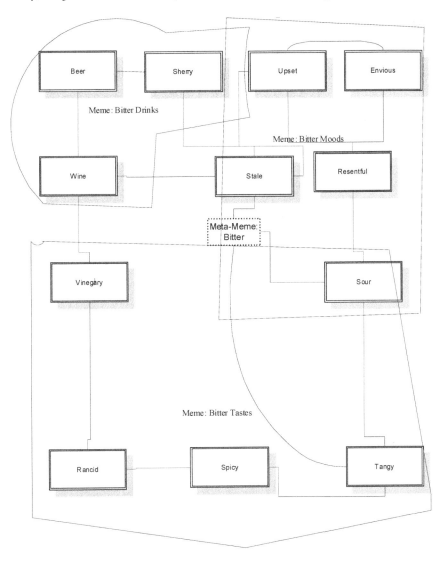

Figure 1. Schematic/Memetic Information Storage.

Source: Author.

Survival entails quickly abandoning an old pattern of schema organization and learning innovatively how to adapt to the changed environment. This new learning, represented by a new pattern of schematic organization (that is, by a new meme) is then associatively incorporated into the individual's schematically-arranged knowledge base. Adaptive learning in response to unanticipated circumstances — a crisis in the context which I define above, becomes revolutionary.

Crisis occurs when existing patterns of behavior, existing patterns of schema organization, are no longer useful in guaranteeing our survival. When there is some kind of fundamental change in the world, we must either adapt and innovate, or we perish.

This was the case following the end of the last Ice Age. Agriculture, settled life, and then civilization itself resulted from mankind's innovative response to that crisis. Ironically, the solution to that long-ago crisis, civilization, has ultimately led ten thousand years later to today's unsustainable high-energy, hydrocarbon-based civilization. Hydrocarbon-based dependency has, in its turn, led directly to the yawning precipice of our own potentially civilization-ending existential crisis.

To survive this catastrophe, before it's too late, we must change not only how we do things, but at our must fundamental level we must change the way we operate. A change of schema, of meme and indeed of meta-meme, is mandated — if we are to survive, individually and as a civilization — by the changed global reality.

To learn from these crises and adapt, we would need to create a new culture, a new civilization that can endure for ages to come.

We must see that this opportunity is not squandered.

CHAPTER 2. SYSTEMS THEORY AND ITS IMPLICATIONS

In a system, the chains of consequences extend over time and many areas: The effects of action are always multiple...We can never do merely one thing. Wishing to kill insects, we may put an end to the singing of birds... — Robert Jervis[19]

A SYSTEMIC PERSPECTIVE

Reality is a nested system of systems. Nature, the biosphere, human civilization, as well as civilization's political and economic components, are all systems. Nature is itself a system that contains, nested within it, all human systems. And these are themselves contained within and constrained by, the wider natural systems of Earth and cosmos. Understanding systems is essential if we are to understand our present day reality.

There are human systems and there are non-human systems. The focus in this book is on human systems, and how their outputs affect our surrounding non-human systems.

Understanding the basic properties of systems is essential for understanding both why there is order in the world and, more topically, why and how our present day actions are becoming so disruptive of that order.

When we speak of such vast aggregates as "the global economy" or the "international system," or even of individual nations such as the United States, we are speaking about human systems. Large-scale human systems are hierarchical because they contain multiple levels of organization. For example, the international political system is composed of nation-states that interact with one another. However, these elements, the nation states, are themselves com-

---

19. Jervis, Robert, *System Effects: Complexity in Political and Social Life*, Princeton University Press, Princeton, N.J., 1997, pp. 10.

posed of lower level elements, ending ultimately with their citizens; individual human beings.

All social systems are ultimately composed of interactions among people. Of course, the characteristics of humans, principally how our minds store and process information, is foundational to all of these higher level social systems. As we've seen, that storage is associative, schematic, and memetic.

Properties of Systems

All systems possess three universal properties:

1) *Interconnectedness.* A set of units or elements interconnected, so that changes in some elements or their relations produce changes in other parts of the system. Any change to any part of a system causes rippling changes throughout every part of a system.

2) *Emergence.* The entire system exhibits properties and behaviors that are *different* from (and not reducible to) those of its constituent parts. For example, water has the property of wetness, which is not reducible to its constituent molecules of hydrogen and oxygen atoms, but is a property of the whole molecular system. Similarly, conscious awareness is not reducible to any individual neurons in a human brain, but is a collective property of the brain as a whole. Emergence is one of the great miracles of nature.

3) *Boundedness.* All systems have definite boundaries. There is an area that lies within a system's boundary, and an area that lies outside of its boundary.

There are two broad categories of systems:

Simple systems. Simple systems respond mechanically to inputs from their environment and do not learn or manifest other complex behaviors.

Complex adaptive systems. There is no universally accepted definition for a complex adaptive system. A particularly succinct definition is: "A complex adaptive system can be defined as an adaptive network exhibiting aggregate properties that emerge from the local interaction among many agents mutually constituting their own environment."[20] Still another definition of complex adaptive systems involves, "...a medium-sized number of intelligent, adaptive agents acting on the basis of local information."[21]

Human systems, composed as they are of conscious, self-aware actors (people), constitute a type of complex adaptive system. The basic behaviors of humans are essentially similar to those of other living things which dwell on this

20. Tetlock, P.E, & Belkin, A., Counterfactual Thought Experiments in World Politics: Logical, Methodological, and Psychological Perspectives, Princeton University Press, Princeton, NJ, 1996, pp. 260.
21. Casti, John L., Complexification: Explaining a Paradoxical World Through the Science of Surprise, Harper Collins, New York, NY, 1996, pp. x.

Earth: we cooperate, we fight, we mate, and we form organized communities. As do, for example, ants.

However, there is one fundamental behavioral difference between us and the rest of nature's creatures, which is itself an emergent property of our neural complexity: the ability to form schematically coordinated groups by means of language-mediated memetic exchanges, which are capable of inter-coordinating complex activities among and between themselves, over extended periods of time and across great, even worldwide, distances. Recall Chapter 1: "Memes are capable of undergoing evolution according to natural selection. For any environment some are more "fit" than are others. All ideas and beliefs are not equally valid with respect to fitness. Meme complexes can themselves be considered to be complex adaptive systems."

It is this ability to weave meme complexes which coordinate activity among humans and between humans and the natural world which has allowed for the creation of a hydrocarbon-based global economy. For a given amount of energy, the amount of wealth that can be produced depends upon several factors, including the efficiency in which the energy is applied to the production of food or goods, and less obviously, the ability of the system in which production occurs to coordinate the overall production and distribution of goods in an optimal manner. In other words, its productive and distributive efficiency depends upon the relative fitness of its underlying memetic complex adaptive system.

STABILITY AND CHANGE WITHIN SYSTEMS

Systems can change rapidly or remain broadly unchanged across time. Systems that remain unchanged over time are said to be stable, while those that change appreciably over time are unstable. When used in discussing systems theory, "stable" and "unstable" have specific meanings that might not be consistent with popular usage.

A system is said to be stable if it is able to maintain its existing configuration without significant change over time. For example the earth's climate can be said to be stable if its behavior remains consistent with its past behaviors. This allows us to predict its future values by knowing its corresponding past behaviors. A stable system is one that is not changing appreciably.

Because all systems are composed of interconnected parts, and most are nested within other systems, changes in any one part of a system or in inputs to a system from outside of it ripple through the entire system. Changes propagating through a system can be characterized as being either negative or positive. These terms have technical meanings which may not necessarily coincide with popular usage.

Negative feedback is stabilizing for a system, while positive feedback is amplifying. The core idea is that a system is stable if it is controlled by negative feedback that keeps essential variables within prescribed limits, and that it is unstable if it is dominated by positive feedback loops that amplify changes.

The three fundamental properties that all systems share, interconnectedness, emergence and boundedness, are holistically interwoven. Thus, interconnectedness leads to effects rippling across the system. This dynamic interaction gives the system collective or group properties and leads to the phenomena of emergence. Neither of these two effects could occur if the system were not bounded. This is because the existence of a definite boundary means that all paths within a system must form closed loops. Thus chains of causation become circular as opposed to linear. Crucial to our understanding of systemic effects is the concept of feedback. Systems theorist Robert Jervis notes that:

> Feedbacks are central to the ways systems behave. A change in an element or relationship often alters others, which in turn affects the original one. We are then dealing with cycles in which causation is mutual, or circular, rather than one-way, as it is in most of our theories. As with the interaction processes discussed earlier, it is difficult for observers to assign responsibility and for actors to break out of, reinforcing patterns that seem to come from everywhere and nowhere. The actors' behavior collectively causes and explains itself.[22]

He continues:

> Feedback is positive or self-amplifying (and destabilizing) when a change in one direction sets in motion reinforcing pressures that produce further change in the same direction; negative or dampening (and stabilizing) when the change triggers forces that counteract the initial change and return the system to something like its original position. Were it not for negative feedback, there would be no stability as patterns would not last long enough to permit organized society. Without positive feedback, there could be no change and growth.[23]

As a concrete example, human emissions of greenhouse gases leads to a general warming of the earth's surface. This warming causes melting of frozen bogs and tundra swamps in the sub-arctic, releasing large amounts of methane into the atmosphere. Methane is a greenhouse gas that is about 20 times as efficient at trapping heat as is carbon dioxide. Thus, human-caused warming unleashes natural greenhouse gases which accelerate the process of global warming. This is an example of positive (amplifying) feedback. At the same time, increased temperatures cause more water vapor to be evaporated into the air. While water vapor traps heat, it also blocks sunlight from reaching the surface of the earth by making the atmosphere hazy or cloudy. This latter process decreases temperatures, which counteract general warming. This has the effect

---

22. Ibid., Jervis, pp. 125, #19.
23. Ibid.

24

of stabilizing temperatures. Here a negative (stabilizing) feedback loop is at work.

Reality is complex; many negative and positive feedback loops are all operating at the same time.

COMPARATIVE ADVANTAGE

The 21st century global economy is predicated upon international trade. Trade is one aspect of present day civilization which generally goes unnoticed and unremarked — we simply take it for granted. This is unfortunate because it gives us important insights not only into how we have become as wealthy as we have and, concomitantly, as powerfully disruptive to nature as we have — but because it also offers insights into the deeper organizational forces of nature itself. It offers us a post-catastrophe tool that properly used can help to facilitate the emergence of a new civilization or civilizations.

Warfare is at best a zero-sum game. In war, if country A conquers country B, then country A may, at most, gain an amount of wealth for itself from country B, equal to or (due to destruction) less than what country B loses.

Trade, however, is a non-zero sum game. This is because of a phenomenon codified in Ricardo's Law of Comparative Advantage. Stated as generally as possible, it asserts that when two (or more) groups, each of which produces various goods and/or services with differing degrees of efficiency (cost), trade together, the trade can be to their mutual advantage, because each player will have an incentive to focus on what he can produce most efficiently. This is explained further in Chapter 9.

Within-group cooperation, in conjunction with between-group conflict, is an ancient resource allocation mechanism for all living organisms. But within-group cooperation, in conjunction with between-group cooperation, as exemplified by trade, is a unique discovery of humanity.

Human civilization has stumbled upon an emergent, organizing property of nature itself, which when applied to the human economy allows for greater wealth creation with limited material and energy resources. This is because between-group cooperation allows for greater energy efficiency, leading to greater wealth production than is possible otherwise. This is of course a good thing — in principle. But like all good things, and like all new tools and techniques, it must be used wisely. As such it represents the crossing of a threshold of organizational complexity with respect to its animating adaptive meme complex which facilitates the emergence of hitherto untapped opportunities, potentials — and dangers.

The advantages of between-group cooperation come from tapping into the systemic principle of emergence. But there are problems — big problems — that come about because of the inefficient manner, usually designated as "capitalism" or as a "market economy," in which this principle is actually utilized by modern civilization.

Trade is based upon the desire to gain personal wealth. The process of gaining wealth is consciously understood to be based upon individual greed, and is considered to be amoral. Supposedly, Adam Smith's "invisible hand" renders the aggregate of individual self-enriching actions of private gain into a universal public good. Individuals, therefore, need not consider the consequences of their actions upon the wider social and biospheric systems in which their actions take place. Because human attention is biologically limited to one thing at a time, it is very easy to simply disregard the potential effects upon both human society and nature itself, of what we do to produce our individual gain.

This means that human systems have discovered a principle of systemic organization that is unknown in our surrounding living (biospheric) and natural systems (weather). The leverage afforded by exploiting this between-groups organizing principle gives the human system great power to affect not only its own state, it also gives it the ability to destabilize our surrounding living and natural systems upon which our very existence depends. Global industrialization can cause a mass extinction of other life forms, and it can substantially change the climate. As with all "tools," this one can be abused. And it is being abused — mightily.

Paul Ormerod is a prominent researcher and London-based economist.

> For Ormerod, there may be very rare but similar qualitative leaps in the organisation of society. The creation of cities, he believes, is one. Cities emerged perhaps 10,000 years ago, not long after humanity ceased being hunter-gatherers and became farmers. Other apparently progressive developments cannot compete. The Roman Empire, for example, once seemed eternal, bringing progress to the world. But then, one day, it collapsed and died. The question thus becomes: is our liberal-democratic-capitalist way of doing things, like cities, an irreversible improvement in the human condition, or is it like the Roman Empire, a shooting star of wealth and success, soon to be extinguished? Ormerod suspects that capitalism is indeed, like cities, a lasting change in the human condition. "Immense strides forward have been taken," he says. It may be that, after millennia of striving, we have found the right course. Capitalism may be the Darwinian survivor of a process of natural selection that has seen all other systems fail.[24]

The core of the problem is due to the emergence of undying artificial "persons" whose sole reason for existence is to maximize profits over the short term — that is, corporations. These entities have effectively destroyed any semblance of a true market economy. This is because a true market's productivity is predicated upon voluntary cooperation and specialization within the economic group (the "firm"), in conjunction with voluntary competition on a level playing field between firms producing similar or competing goods.

---

24. *The Sunday Times*, Oct. 16, 2005, Waiting for the lights to go out, http://www.timesonline.co.uk/article/0,,2099-1813695_1,00.html

But once corporations achieve legal personhood, the largest of these entities are able to methodically employ their vast resources to gain effective control over governments, and to thereby rig the game of cooperation and competition in their favor. Smaller competing corporations are either smashed or assimilated via corporate takeover.

Other entities such as unions, interest groups, political parties, and suchlike, are simply unable to check and balance the mammoth global corporations. There are two main reasons for this: (1) corporations are global actors, whereas their opposing groups are usually active only at a national level; and (2) corporations have legal personhood status under the law, while no other entity possesses such status. This asymmetry means that corporations almost always prevail in any contest.

Things are even worse when individual humans attempt to challenge these artificial "persons." The law assumes that both litigants are equally human. Of course the corporations possess resources far beyond those of any mortal person. And so they generally prevail.

Consequently, the creative ferment of true human-to-human cooperation and competition in the context of a level playing field is lost. In fact, all human goals, including any sense of responsibility for other humans and for nature itself, are lost as well.

CORPORATE TAKEOVER OF THE GLOBAL POLITICAL ECONOMY

This abuse of between-group cooperation in the form of a market economy becomes potentially fatal for most of humanity and for civilization itself in modern times, because we live in a human system characterized by its being a high technology, high energy, globalized political economy.

All political economies have two dimensions: a political dimension and an economic one. In our case, economic activity is global in scale and scope. Its fundamental constituent elements are the trans-national corporations. Political activity is geographically constrained to the boundaries of the world's nation states. There is no global political entity which possesses the kind of police powers that states do within their own borders.

The effect of this over time has been to put the geographically unbounded trans-national corporations in effective control of the governments of the world's nation-states. An international system has emerged which is ordered around the rigged exploitation of Ricardo's Law of Comparative Advantage for the individual gain of the few who control the trans-nationals, and not for the laboring many.

These trans-national corporations are systems in their own right. They consist of humans grouped into task-oriented departments. These departments exhibit a maximal degree of within group cooperation. They are bounded: everything outside of the corporation is of no concern to the corporation except insofar as it aids or hinders the achievement of the corporations profit seeking agenda. A collective behavior emerges from this highly structured, rule-based human system. The corporation becomes a thing in itself, separate and distinct from any of the human elements which animate it. In plainer words, corporations are material-gain seeking systems, whose constituent elements are humans. It should also be noted that these entities are not "alien" invaders assailing the global political economy. They were created by human beings to facilitate the human goal of rapid personal enrichment. They therefore represent the reification of our values. We have no one to blame for their actions except ourselves.

The wealth creation powers of inter-group cooperation are thus subverted to facilitate the gains of the few, regardless of any detriment to the toiling many. The effects of the corporation's wealth-creating activities (industry) upon nature (pollution) are disregarded entirely. If citizens have the temerity to attempt to use political means to constrain the harmful effects of corporate wealth creation, the corporation simply uses its vast resources to ensure governmental non-interference. And in the United States, corporations are fictitious "persons" entitled to the same Constitutional protections as are real persons. Across the world, the people no longer control their governments — rather these fictitious "people," the corporations, do.

Indeed from the origins of civilization until the present day, no human-created entities have ever been as wealthy, and concomitantly as powerful across the globe, as are today's hydrocarbon energy companies. The corporate elites who control these companies control the relevant policies of the world's nation states. Their sole goal is to maximize short-term individual gain for the wealthy few. In fact, it is illegal for a corporation to do anything else. The bottom line, quarter by quarter, is truly the bottom line! On June 8th, 2005 many media outlets reported this wire story:

> WASHINGTON (Reuters) — A White House official, who previously worked for the American Petroleum Institute, has repeatedly edited government climate reports in a way that downplays links between greenhouse gas emissions and global warming, The New York Times reported on Wednesday. Philip Cooney, chief of staff for the White House Council on Environmental Quality, made changes to descriptions of climate research that had already been approved by government scientists and their supervisors, the newspaper said, citing internal documents...[25]

It is becoming increasingly rare for such acts of malfeasance to even be reported by the mega-corporations that provide us with "news." Increasingly,

---

25. Tiscali.news, Aug, 6, 2005, Official Edited Warming and Greenhouse Emission links-Report, http://www.tiscali.co.uk/news/newswire.php/news/reuters/2005/06/08/world/officialeditedwarmingandgreenhouseemissionlinks-report.html

the actions of political decision-makers are seamlessly melded to those of the corporate elites.

As hydrocarbon energy becomes scarcer, the laws of supply and demand mandate that it will become more expensive. The more we pay, the richer the plutocracy becomes. The sad fact that global economy rests upon an unstable, unsustainable foundation of diminishing hydrocarbon energy is simply shrugged off; higher prices mean ever growing wealth for these people. Another news article published that same day was even more telling:

> An unprecedented joint statement issued by the leading scientific academies of the world has called on the G8 governments to take urgent action to avert a global catastrophe caused by climate change. The national academies of science for all the G8 countries, along with those of Brazil, India and China, have warned that governments must no longer procrastinate on what is widely seen as the greatest danger facing humanity. The statement, which has taken months to finalise, is all the more important as it is signed by Bruce Alberts, president of the US National Academy of Sciences, which has warned George Bush about the dangers of ignoring the threat posed by global warming.[26]

We are warned about the need to take, "urgent action to avert a global catastrophe caused by climate change." Who is issuing this alarmist appeal — the usual doomsday nuts, perhaps? No, "the leading scientific academies of the world."

Considering that the US government failed to act on this appeal, it is quite apparent who controls it. It is decidedly not the citizens. The corporations rule in the United States. When the Chief of Staff for the White House Council on Environmental Quality is a former executive from the American Petroleum Institute, it is clear that quarterly profits above all else will prevail.

However, the desire for unlimited material gains in the short-term led to the development of a nested subsystem for economic wealth production — the corporation — which came to dominate the entire human system. Now endless growth to achieve short-term material gain with no regard for anything else has become the aim of human civilization. Think of this as memetic cancer wherein the "cells" proliferate rapidly throughout the body with no objective except to proliferate further. The result for both an individual body as well as for the body of human civilization — for the biosphere itself — is identical in both cases — death.

And so, like the astronauts onboard the ill-fated space shuttle Columbia, the lights are on, the air is circulating; everything seems fine — except that we are on a trajectory that leads inexorably towards catastrophe.

---

26. http://news.independent.co.uk/world/science_technology/story.jsp?story=645071

TRAJECTORY OF HUMAN CIVILIZATION

Human civilizations are complex adaptive memetic systems which have trajectories through space and time. These trajectories are not random but rather are largely internally generated. In other words, what we do affects where we go. And what we think, how we think and how we perceive reality determines where we as a global civilization go.

Civilization's historical trajectory is ultimately decided by some combination of external variables such as changes in climate that are not due to human activity (volcanic eruptions for example) and, more importantly, by internally-generated, human-caused factors — such as global warming, peak oil, and depletion of fresh water supplies. Humanity's responses to these crises then determine the future trajectory of civilization. The impending crises are not externally-caused but rather are internally generated.

We, and ultimately our very beliefs and values, are the cause of these crises. Accordingly, reacting to the brewing crises absolutely requires us to fundamentally alter our values. There is no such prerequisite for dealing with naturally occurring, externally-generated crises.

Human civilization is moving through a changing environment. The things that we do and the ways that we organize ourselves shape and alter this environment. This environment is filled with "strange attractors." Strange attractors were first discovered mathematically by scientists investigating chaos theory. Here is a simple definition:

> In dynamical systems, an attractor is a set to which the system evolves after a long enough time. Once trajectories get close enough to an attractor, they remain close even if slightly disturbed. An attractor can be a point, a periodic trajectory, a continuous manifold that gives rise to a trajectory that is periodic or aperiodic, or even a complicated set with fractal structures known as a *strange attractor*. Describing the attractors of chaotic dynamical systems has been one of the achievements of chaos theory.[27]

Think of bowling, and that teetering gutter ball that just cannot escape once it reaches the edge of the gutter, or a marble spiraling into a depression when it rolls too close to the downward-sloping sides. These attractors represent "places" towards which human civilization is are drawn once we approach them closely. These attractors represent outcomes of historical trends which become inevitable after a certain point.

The set of crises toward which civilization is now being drawn represents just such an attractor. No one knows for sure just when we will run out of oil or fresh water, but it is getting harder and harder to avoid such a calamity.

---

27. *Encyclopædia Britannica*, 2006, Chaos, Encyclopædia Britannica Premium Service, http://www.britannica.com/eb/article-9022470.

Civilization was given a new historical trajectory when mankind learned to use agriculture to sustain ever-larger populations. These larger populations developed ever more complex and coordinated ways to create wealth from nature. By discovering the principle of comparative advantage, they were able to do this with ever greater efficiency. By discovering and exploiting the vast energy bonanza provided by hydrocarbon fuels, in the context of a corporate-controlled political system, civilization moved inexorably to a trajectory leading towards an attractor which represents the onset of multiple crises which potentially can end civilization — at least as we have known it.

The "space" itself, technically termed a "fitness landscape," can be considered to be a learning matrix, because as the global system is compelled by crisis to "learn," to anticipate, and to cope with crises, it reconfigures itself. Old attractors disappear and new ones form — destinies change.

The figures on the facing page illustrate visually the historical trajectory of civilization through an attractor-filled learning space.

Keep in mind that humans are the ultimate elements forming this ultimate human system, global civilization. As humanity changes, the very nature of the system changes — everything affects everything else in a system. These changes can and do alter the systemic trajectory.

Changing the way people think and act in a coordinated, systemic way can change the future of civilization. And if such a comprehensive change of memes could be achieved, perhaps a rosier future might yet be a possibility.

The peak oil attractor and its associated crises, including global warming, provide the opportunity for comprehensive schematic/memetic change. A civilized, humane future which is sustainable for endless ages is a possible outcome. However, in the here and now, it is hard to see where such healthy coordinated, systemic changes might come from; citizens no longer control their own governments. Impersonal, short-term profit-driven, theoretically immortal artificial persons better known as "corporations" do. Because of this takeover, amplifying feedback loops have been created that are undermining all of our "life support" systems and that would appear to make the continued existence of the contemporary human system impossible.

At the very center of it all are the ordered patterns of memes, schematically encoded in the inter-neuronal connections of our physical brains from which, in turn, our minds emerge. The resulting complex adaptive group memetic system facilitating group interaction and coordination of behavior and understanding of meaning among and between large numbers of human beings determines the ordering of human civilization, and hence how it interacts with our surrounding biological and natural systems. Caring, cooperative, social, political, economic, and societal evolution can be facilitated by new understandings of reality evolving upon a much transformed fitness landscape. The meme complexes of civilization can evolve to become more "fit."

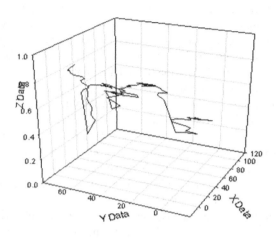

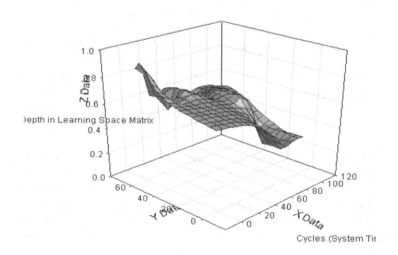

Figure 1. *Sample Trajectory of Human Civilization in a Learning Space Matrix.*
Source: Author.

PART II
THE CRISES

CHAPTER 3. PEAK OIL AND HYDROCARBON RESOURCE DEPLETION

Understanding depletion is simple. Think of an Irish pub. The glass starts full and ends empty. There are only so many more drinks to closing time. It's the same with oil. We have to find the bar before we can drink what's in it.
— Dr. Colin J Campbell[28]

PEAK OVERVIEW

The world is positioned on the brink of a historical chasm. Everything familiar is about to vanish — beginning with where and how most of us live. The physical and proximate cause of this wrenching change is the imminence of what is called peak oil — the point at which half of all of the oil the world holds will have been taken from the ground. From that point onwards, every year less oil will be produced, at ever greater cost.

Civilization is almost completely reliant upon growing supplies of cheap oil. But oil is a non-renewable resource, at least within the human timescale. For the planet as a whole, the amount of oil discovered each year reached a peak about forty years ago and has been declining ever since.

The oil supply curve is called the "Hubbert Curve," because it was first described back in the 1950s by legendary petroleum geologist M. King Hubbert. So long as there was more oil in the ground waiting to be discovered than was consumed, the supply of cheap oil was ever-growing. But once the peak of oil production is hit, that will be the end of cheap oil. The effects of this will be world shaking.

---

28. Peak Oil.net, Colin Campbell, http://www.peakoil.net/Colin.html

Building a way of life around cheap oil, the US and an ever-growing plurality of the developed world began to center life on the "suburban dream," where each small nuclear family lived in its own home in a densely residential area located a considerable distance from the parents' places of work. Many families have at least two cars. They drive to school, to work, and to big box stores located at the margins of town. The same pattern is true of entertainment and necessities of life such as medical services. Even in China, with its burgeoning industrial manufacturing economy, ambitious people aspire to some version of this.

But this model is grossly inefficient, and cannot be replicated for Earth's six billion inhabitants.

Nature is not cruel; nature is neutral. It is utterly indifferent to the desires and the delusions of humanity. The suburban dream economy is based ultimately upon a foundation of cheap hydrocarbon energy, primarily oil, natural gas (methane), and coal, in essentially limitless quantities — and that is not reality. It's also based upon the notion that the earth's biosphere can absorb any amount of human industrial activity while still keeping the climate stable — which it cannot.

As Herbert Stein, a former Chairman of the Council of Economic Advisers under Presidents Nixon and Ford once said, "Things that can't go on forever don't."[29]

PEAK OIL: PHYSICAL REALITIES

Oil and natural gas are formed by the compression and heating of dead plant and animal matter crushed beneath tons of soil, stone, and seawater over hundreds of millions of years. Coal is formed by similar processes on ancient swamp plants. For human purposes, these resources are non-renewable because they require many millions of years to form.

In 1956 a well-respected petroleum geologist named M. King Hubbert predicted, using a mathematical modeling technique called the logistic decline model,[30] that petroleum production in the US would peak in 1970 and that for the world as a whole, the peak year of production would be 2000. This seemed heretical at the time but, in fact, for the US the actual peak in oil production did occur in 1971 and many geologists, petroleum scientists, and industry analysts have come to accept that Hubbert got it right. Because the 1973 and 1979 oil

---

29. Krugman, Paul, This can't go on, Nov. 4, 2003, *New York Times*, http://truthout.org/docs_03/110503J.shtml

30. Sandrea, Rafael, What about Deffeys' prediction that oil will peak in 2005?, http://www.gasandoil.com/goc/features/fex53910.htm

shocks actually reduced demand for oil briefly, the projected peak in global production was delayed until just about now, in the first decade of this century.

Somewhere between about 2005 and 2012, global production of oil will reach its all-time peak.[31] This event is universally referred to as "peak oil." From that point onwards, production will decline inexorably. According to Princeton geologist and peak oil researcher Dr. Kenneth Deffeyes, author of the informative book *Beyond Oil: The View from Hubbert's Peak*, the global peak of oil production actually occurred on December 16, 2005:

> The world peak would then happen when 1.0065 trillion barrels have been produced (half of 2.013). Following Hubbert, I used the Oil & Gas Journal end-of-year production numbers. It isn't that the Oil & Gas Journal reports are divinely inspired; their methodology is well explained and their reports constitute a relatively consistent data set. The cumulative world production at the end of 2004 was 0.9812 trillion barrels and at the end of 2005 it was 1.00748 trillion. During the year, we passed the halfway point. The graph shows the date of the crossover: December 16, 2005.[32]

And at the same time that production of oil is reaching its peak, world oil consumption is increasing by about two to three percent per year. The rapid industrialization of China, followed closely by other nations such as India, is further accelerating the increase.

Natural gas is following a similar production trend, with its peak occurring before 2020 according to most of these same experts.[33]

Research by Stuart Staniford, PhD, a physicist, makes a convincing case that Saudi Arabia's claimed oil reserves are massively overstated. Staniford concludes on the basis of careful statistical analysis of oil industry data that the Saudis have actually produced 105 billion barrels out of 180 billion barrels or so, and are at around 55%-60% of their ultimate recovery. In short, Saudi Arabia's oil output is seen to be well past its peak, sustained only by massive use of increasingly advanced technologies, and is approaching a sudden and precipitous decline.[34] Only Herculean efforts using advanced oilfield recovery technologies have kept Saudi output relatively stable in recent years.

Staniford reaches similar conclusions regarding Kuwait's reserves, finding that current cumulative production is 36 billion barrels, with about 40 billion barrels remaining.

---

31. C.J. Campbell, Presentation at the Technical University of Clausthal, Dec. 2000, http://www.geologie.tu-clausthal.de/Campbell/lecture.html

32. Beyond Oil, The View From Hubbert's Peak, Feb. 11, 2006, http://www.princeton.edu/hubbert/current-events.html

33. Ibid., C.J. Campbell, # 31

34. Saniford, Stuart, What should we have predicted about Kuwait? http://www.theoildrum.com/story/2006/1/20/193723/259#more

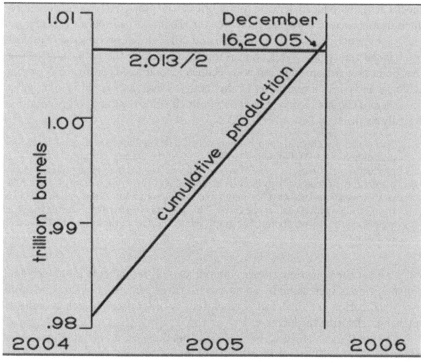

Figure 1. Peak Oil Date.

Source: Prof. Kenneth Deffeyes.[35]

These conclusions mean that Saudi Arabia's production capacity is in terminal decline, while Kuwait's oil production capacity will soon begin to decline. Staniford believes that there will be sharp declines in Saudi output and slower decline rates, beginning slightly later, for the Kuwaiti oilfields.

This finding is wholly consistent with those of oil industry insider Matt Simmons, an oil industry investment banker and a member of the 2001 Bush-Cheney Energy task force. In his massively researched study of Saudi Arabian oil reserves, published as *Twilight in the Desert: the Coming Saudi Oil Shock and the World Economy*, Simmons reaches the same overall conclusions. Regarding the role of advanced recovery technologies in maintaining oil output, Simmons observes that:

> None of these technical breakthroughs created an "oilfield fountain of youth," which is what would be required for the [optimistic] forecaster's scenarios [of endless future production] to unfold. Instead, these advances combined to extract the easily recoverable oil from giant fields even faster, and led to decline curves, once

35. Prof. Kenneth Deffeyes, www.princeton.edu/hubbert/

high reservoir pressures depleted, steeper than the industry had ever experienced before. Thus, I must conclude that the odds are better than even that oil output from at least several of Saudi Arabia's key oilfields is now at risk of entering an irreversible decline. The odds are only marginally lower that output might soon decline in every one of its old fields. Moreover, the "new" rehabilitation projects for Saudi Arabia's troublesome mothballed fields might not work out as successfully as planned. All of these eventualities put the optimistic forecasts at grave risk of remaining unfulfilled. It is likely that they are wrong.[36]

The effects of advanced oilfield recovery technologies can be seen very clearly in this graph which is taken from a paper written by two researchers from the Rensselaer Polytechnic Institute, John Gowdy and Roxana Julia, entitled "Technology and Petroleum Exhaustion: Evidence from Two Mega-Oilfields." The significance of this is that the rate of decline in oil production, post-peak, will be significantly greater than most researchers, and most models, assume.

Thus the window of opportunity to deal with this set of problems is already almost closed. This mitigates, as we shall see, in favor of military approaches to dealing with this set of problems, over technological ones.

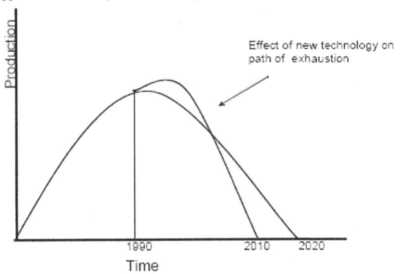

Figure 2. Technological Advance Masks Impending Production Declines.
Source: Rensselaer Polytechnic Institute.[37]

---

36. Simmons, Matthew R, Twilight in the Desert: The Coming Saudi Oil Shock and the World Economy, John Wiley & Sons Inc, Hoboken, NJ, 2005.
37. Rensselaer Polytechnic Institute, Dec. 2005, Revised April 2006, Working papers in Economics, Technology and Petroleum Exhaustion: Evidence from Two Mega-Oilfields, http://www.economics.rpi.edu/workingpapers/rpi0512.pdf.

Demand for oil is poised to permanently exceed its available supply. The inevitable consequence of this is rapidly increasing energy prices. At first prices will be very volatile. With supply and demand closely matched, even minor disruptions in supply, such as a fire at a refinery, a strike among oilfield workers, or political unrest in an oil-producing country, will trigger sudden and rapid price spikes. Prices will eventually come down somewhat, but the overall trend will be ever upwards.

Published news reports quote Matt Simmons as taking an unambiguous view on what peak oil means in terms of near future oil prices:

> The former energy advisor of US President George W. Bush, Matthew Simmons, predicts oil prices could reach as much as 250 US dollars per barrel over the coming years. "We have to expect an oil price between 200 and 250 dollars per barrel," Simmons was quoted as saying in the January issue of the German-based *Capital* economic magazine. He cited an imminent shortage of oil supply and a growing global demand, especially in China and India, for the sharp rise in oil prices. Simmons who heads the American energy investment firm Simmons & Co., pointed also to a worldwide decline in conventional oil discoveries in new fields. He anticipated global oil output could drop from the current 75 million to 65 million barrels per day by 2012.[38]

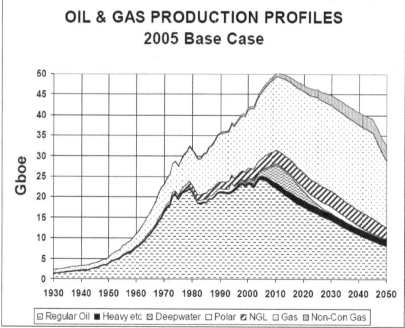

Figure 3. Oil and Gas Liquids 2004 Scenario.

Source: Colin J. Campbell/Association for the Study of Peak Oil and Gas Newsletter[39]

38. *Islamic Republic News Agency*, Jan. 3, 2006, Bush's former energy advisor expects oil to hit $250.00 per barrel, http://www.irna.ir/en/news/view/menu-234/0601035729190447.htm

Discoveries of new oil reserves peaked in the 1960s and have been declining ever since (see figure 3 below). According to the latest statistics, as of January 2006, for every five barrels of oil consumed, only one is discovered.[40]

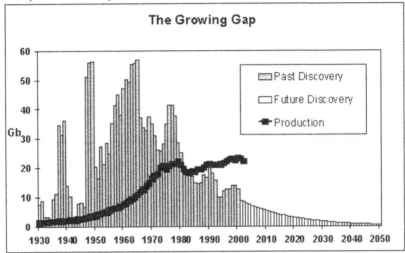

Figure 4. The Growing Gap.

Source: Colin J. Campbell[41]

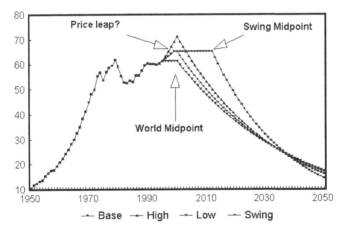

Figure 5. World Oil Production: 1950-2050.

Source: Colin J. Campbell[42]

39. Colin J. Campbell, Association for the Study of Peak Oil and Gas Newsletter, No. 64, April 2006, http://www.peakoil.ie/downloads/newsletters/newsletter64_200604.pdf
40. Association for the Study of Peak Oil and Gas, Ireland, Jan. 2006, The questionable contribution of enhanced recovery, http://www.peakoil.ie/newsletters/710
41. Colin J. Campbell, Hubert Peak.com, http://www.hubbertpeak.com/campbell/

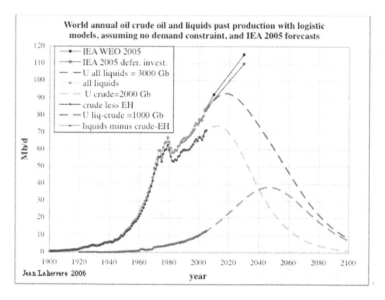

Figure 6. World Hydrocarbon Liquids Production: 1900-2100.

Source: Jean Laherrere[43]

These two divergent trends — ever growing demand and ever diminishing supply — indicate that essentially all of the planet's recoverable oil reserves will have been consumed by mid-century. However, the problems will have become acute long before then. A car runs fine until the moment that it burns the last drop of gas in its gas tank. Not so the high-energy global civilization.

With local, national and global economies based upon an assumption of endless supplies of cheap oil, persistent deepening recession is an inevitable near-future consequence.

In 2005 researcher Robert Hirsch produced a study that was commissioned by the US Department of Energy. In the opening paragraph of the study's Executive Summary, he noted:

> The peaking of world oil productions presents the US and the world with an unprecedented risk management problem. As peaking is approached, liquid fuel prices and price volatility will increase dramatically, and, without timely mitigation, the economic, social, and political costs will be unprecedented. Viable mitigation options exist on both the supply and demand sides, but to have substantial impact, they must be initiated more than a decade in advance of peaking.[44]

42. http://www.oilcrisis.com/campbell/camfutur.htm

43. *Modelling future liquids production from extrapolation of the past and from ultimates,* Jean Laherrere, Paper presented at the International Workshop on Oil Depletion, Uppsala University, Sweden, May 23-24th, 2002, *http://www4.tsl.uu.se/isv/IWOOD2002/ppt/UppsalaJHL.doc*

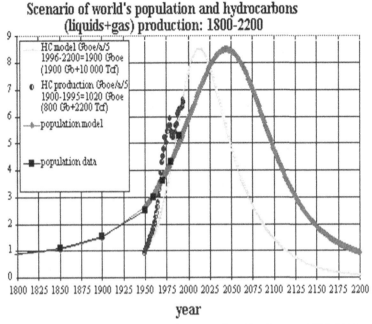

Figure 7. World Population and Hydrocarbon Production: 1800-2200

Source: Source: Jean Laherrere[45]

There is no large-scale substitute for hydrocarbon energy, as will be discussed later in this chapter. Even if there were, there is now no time to convert the world's several billion vehicles to a new energy source. The growth in demand for oil is now nearly three percent compounded per year. Time is running out to deal with the economic ramifications of peak oil.

Cal Tech Physicist David Goodstein[46] carefully analyzed the energy options and concluded that a decade is almost certainly insufficient time to adapt:

---

44. *Peaking of World Oil Production: Impacts, Mitigation, and Risk Management*, Robert L. Hirsch SAIC et al., Feb. 2005, http://www.netl.doe.gov/publications/others/pdf/ Oil_Peaking_NETL.pdf

45. *Future sources of crude oil supply and quality considerations*, J. H. Laherrere, Paper presented at DRI/McGraw-Hill/French Petroleum Institute Conference "Oil markets over the next two decades: surplus or shortage?" Rueil-Malmaison, France, 12-13 June, 1997, http://www.oilcrisis.com/laherrere/supply.htm

46. Professor Goodstein has a very informative 20 minute video interview dealing with peak oil related issues online at: http://etopiamedia.net/empnn/pages/cpt-emnn/cpt-emnn599-5551212.html.

ESTIMATED PRODUCTION TO 2075										End 2005	
Amount			Gb	Annual Rate - Regular Oil						Gb	Peak
Regular Oil				Mb/d	2005	2010	2015	2020	2050	Total	Date
Past	Future		Total	US-48	3.6	2.8	2.2	1.7	0.4	200	1971
Known Fields	New			Europe	5.0	3.4	2.3	1.6	0.2	75	2000
968	794	138	1900	Russia	9.2	8.5	6.9	5.7	1.5	220	1987
	932			ME Gulf	20	19	19	19	11	680	1974
All Liquids				Other	29	27	23	20	9	725	2004
1073	1377		2450	World	67	61	54	48	22	1900	2005
2005 Base Scenario				Annual Rate - Other							
M.East producing at capacity				Heavy etc.	2.3	3	4	4	4	151	2021
(anomalous reporting corrected)				Deepwater	3.6	12	11	6	4	69	2011
Regular Oil excludes oil from				Polar	0.9	1	1	2	0	52	2030
coal, shale, bitumen, heavy,				Gas Liquid	6.9	9	9	10	8	276	2035
deepwater, polar & gasfield NGL				Rounding				-1	-2	2	
Revised 03/03/2006				ALL	80	86	80	70	37	2450	2010

Figure 8. Table of Estimated World Hydrocarbon Liquids Production to 2075.

Source: Colin J. Campbell/Association for the Study of Peak Oil and Gas Newsletter[47]

Once Hubbert's peak is reached and oil supplies start to decline, how fast will the gap grow between supply and demand? That is a crucial question, and one that is almost impossible to answer with confidence. We can make a crude attempt at guessing the answer as follows: The upward trend at which the demand for oil has been growing amounts to an increase of a few percent per year. On the other side of the peak, we can guess that the available supply will decline at about the same rate, while the demand continues to grow at that rate. The gap, then, would increase at about, say, 5% per year. That means that, 10 years after the peak, we would have to have a substitute for close to half the oil we use today, something approaching 10-15 billion barrels per year. Even in the absence of any major disruptions caused by the oil shortages after the peak, it is very difficult to see how that can possibly be accomplished.[48]

Back in the 1970s, President Carter tried to warn of the coming conse-quences of oil depletion. He recognized that Americans would have to sacrifice in order to deal effectively with the problem. He was voted out of office. Signifi-cantly, his successor had Carter's newly-installed solar panels torn down from the White House roof.

In the 30+ years since, this problem has been largely ignored. During those years, rather than dealing in a methodical and rational way to fend off this set of crises, as the Hirsch Report envisions, the political system was actually captured by the very energy companies which profit most from maximizing the use of hydrocarbon energy and who have the least incentive to transition away from it.

47. Ibid., Colin J. Campbell, # 41

48. Goodstein, David, Energy, Technology and Climate: running out of gas, in *New Dimen-sions in Bioethics*, Yale University Press, New Haven, CT, 2001, http://www.its.caltech.edu/-dg/Essay2.pdf

Basically, in allowing a fortunate few to make money here and now, we have thrown away the sustainable future for all of humanity — forever.

Goodstein concluded his analysis by noting that while it was possible in theory for human civilization to adapt to the looming sudden decline in hydrocarbon energy fuels, such an adaptation did not appear very likely in reality. He states:

> So, technically, scientifically, the means exist to build a civilization that has everything we think we need, without fossil fuels. The future exists. The remaining question is, can we get there? Scientists are supposed to make predictions. Experiment or observation tests the prediction, and the fate of the scientist's theory, acceptance or rejection rides on the outcome. That's how science works. I have a prediction to make. Here it is: *Civilization as we know it will come to an end some time in this century, when the fuel runs out.* This is different from normal scientific predictions in a crucial way. Usually, the scientist hopes that the prediction will prove to be correct, and merely making the prediction does not change the phenomenon in question. In this case I do hope the prediction will be wrong, and I hope that merely making the prediction will help make it become wrong.[49]

PEAK OIL: STRATEGIES

Nations possessing significant military power, such as the US, will pursue a two-track strategy once the scope of the crisis becomes apparent:

1. Use of military power abroad to secure control of essential energy supplies and transit routes.

2. Desperate crash programs to produce oil substitutes domestically using coal, oil shale and oil sands. This will occur in conjunction with rapidly increasing use of nuclear power.

Given the nearly absolute domination of government by the oil industry, efforts to develop renewable alternative energy sources such as wind and solar power have faced an uphill battle and continue to do so.

Both strategies (1) and (2) above will be insufficient to do anything more than slow the rate of decline and create even greater global imbalances and violence between the haves and have-nots.

Strategy (1) is marketed by the corporate oligarchs under the brand name, "The War on Terror." Fear has served the ruling few as a means to not only control the many but to motivate them to act against their own self-interest as well.

This strategy was articulated, not by the two Bush Administrations, not even by the Reagan Administration, but by Zbigniew Brzezinski, National Security Advisor to President Jimmy Carter. Now called the "Carter Doctrine," the President stated during his January 1980 State of the Union address:

---

49. Ibid., Goodstein, David, #48

> Let our position be absolutely clear: An attempt by any outside force to gain control of the Persian Gulf region will be regarded as an assault on the vital interests of the United States of America, and such an assault will be repelled by any means necessary, including military force.[50]

The United States is almost wholly dependent upon unending supplies of cheap oil. Considering that most of the planet's remaining reserves of oil are found in the Persian Gulf region, it is obvious that control of this oil supply is of vital interest to, actually, all developed, developing, or yet-to-be developed nations. But the planet's biggest consumer of oil incidentally also happens to be the sole superpower. The main difference between the Carter and Bush Administrations in this regard is simply honesty — Carter admitted this interest. The Bush Administration denies it.

Figure 9. The Oil Patch.

Source: *Early Warning Report*[51]

Implementing Strategy (1) will result in unending and increasing guerrilla resistance in the lands that we seek to control. Most of these potential targets

---

50. *Encyclopædia Britannica*, 2006, International relations, Encyclopædia Britannica Premium Service, http://www.britannica.com/eb/article-32977.
51. Maybury, Richard. US & World Early Warning Report, Map of Oil Corridor Iraq-Iran, http://www.chaostan.com/mapcorridor.html

are Arab or Moslem. Nearly all are in or adjacent to one fairly compact geographical region. More than 60 percent of the planet's remaining oil is located in the Middle East, specifically the Persian Gulf. The core of this area, centered on portions of Iran, Iraq, and Saudi Arabia, which is often called "the oil corridor," is about the same size as the state of Indiana!

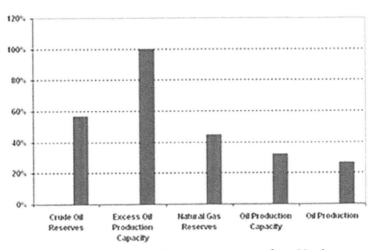

Persian Gulf as a Percent of World (2003)

Figure 10. Persian Gulf Oil as a Percentage of World Oil.

Source: US Energy Information Agency[52]

Whoever controls these oil and natural gas fields and the pipeline routes and sea lanes needed to transport their outputs literally controls the world. Attaining this objective is the hidden goal that lies behind the so-called "War on Terror."

In addition to the raging insurgencies as natives of the region seek to defend their own resource fields (and Iraq and Afghanistan are just the first), fundamental national self-interest will compel major powers such as Russia, and most significantly China, to align militarily against us in order to ensure access to these essential resources for themselves. The maps in this chapter illustrate clearly that the "War on Terror" is in reality a global war for control over the dwindling hydrocarbon energy resources. It manifests itself as a series of wars and occupations of petro-strategic countries such as Iraq, along with geo-strategic ones, for instance those on potential pipeline routes, such as Afghanistan.

52. US Energy Information Agency Country Analysis Briefs, Persian Gulf Oil and Gas Exports Fact Sheet, Sept. 2004, http://www.eia.doe.gov/emeu/cabs/pgulf.html

These are wars which the US and its Western coalition members, alien in culture and religion, coming from far away and thereby possessing extended supply lines, shall inevitably lose. Terrorism does exist; however US military efforts are succeeding only in increasing the numbers of persons willing to resort to this tactic, and mainly to expel us — our military bases and our imposed undemocratic governments — from their home territory. Thus, defeating "terror" cannot be the goal behind actions that increase the probability of terror.

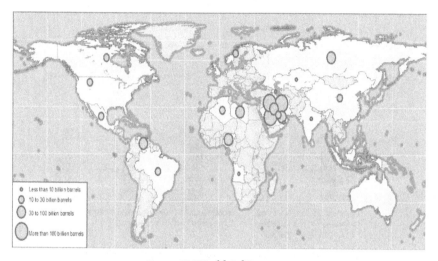

Figure 11. World Oil Reserves.

Source: Prof. Jean-Paul Rodrigue, Hofstra.edu[53]

Obviously, there must be another motive. And there is — energy, and the geopolitics of its control.

However, increasing military resistance to US presence in these geo-strategic and petro-strategic lands and transit routes ensures that we will never be fully able to exploit these vitally needed energy resources. The net result is a double disaster — a military debacle which costs more energy resources than it obtains, coupled with the creation of additional terrorist resistance forces.

It is important to understand that the deep underlying motive for this set of policy choices is an unbounded desire for profit. Multinational energy companies profit immensely from the existing energy order. Accordingly, they employ their effective control of key governments — particularly the American government — to maintain that existing energy order by all possible means including militarily. Consistent with this agenda, they block attempts to implement a rational energy policy based upon transitioning away from hydro-

---

53. http://www.people.hofstra.edu/geotrans/eng/ch5en/appl5en/img/Map_oilreserves.pdf

Major Oil Flows and Chokepoints, 2003

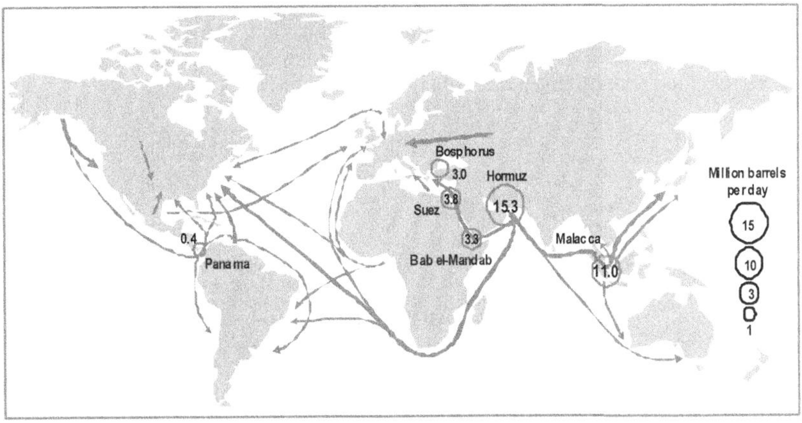

carbon energy and towards renewable energy sources. Still, they must appear to do something. This leads to activating Strategy (2).

Figure 12. Global Oil Flows, Chokepoints, and Transit Routes.

Source: IBID[54]

The looming specter of military and economic disasters will, in the near future, bring about redoubled efforts to develop alternative energy sources. Strategy (2) above will come into play. However, given their political dominance, it will be implemented by the same energy corporations which have already led humanity to the precipice. Vast resources will be wasted on the production of oil substitutes from oil shale, oil sands, and coal.

These efforts, though immensely profitable to the energy multinationals due to governmental tax credits and subsidies, will fail to resolve the underlying problem of energy resource depletion for a simple reason: At best, the amount of net energy produced will be only marginally greater than the amount of energy used to produce the oil substitute. The underlying concept is called "energy returned on energy invested," and is usually abbreviated as EROEI.[55] This is a ratio consisting of the amount of energy required to produce some amount of energy bearing resource. It can be succinctly defined as follows:

54. Hofstra.edu, World Oil Reserves 2003, http://www.people.hofstra.edu/geotrans/eng/ch5en/appl5en/ch5a1en.html
55. Cleveland, Cutler, J., Costanza, Robert, Hall, Charles, A. S., Kauffman, Robert, Energy and the US Economy: A Biophysical Perspective, *Science*, New Series, Vol. 225, No. 4665, (Aug. 31, 1984) 890-897

In physics and energy economics EROEI (energy returned on energy invested) is the ratio between the amount of energy expended to obtain a resource, compared with the amount of energy obtained from that resource. When the EROEI of a resource becomes 1 or less, that energy source becomes an energy sink and can no longer be used as a primary source of energy. For example, when oil was originally discovered, it took on average one barrel of oil to find, extract, and process about 100 barrels of oil. That ratio has steadily declined over the last century to about 3 in the US (and about 10 in Saudi Arabia).[56]

Existing and Potential Oil and Gas Export Routes From the Caspian Basin

Figure 13. Existing and Potential Oil and Gas Export Routes from the Caspian Basin.
Source: US Energy Information Administration[57]

An example of EROEI limiting potential gains from Strategy (2) above is offered by the vast deposits of oil sands which exist in Canada. Transforming these hydrocarbon resources into synthetic oil requires the use of large amounts of another rapidly-depleting hydrocarbon resource: natural gas. Also, huge amounts of water are required. This water is permanently polluted, and can no longer be used for human or animal consumption or for agriculture unless

56. EROEI.com, http://www.eroei,com.
57. US Energy Information Administration, Country Analysis Brief, Caspian Sea, http://www.eia.doe.gov/emeu/cabs/Caspian/Background.html

enormous amounts of energy are employed to detoxify it. The overall result is probably a net energy loss. Although the energy corporations may make fortunes from this effort due to governmental subsidies and tax incentives, it will not come close to solving the energy catastrophe. Even greater difficulties limit utilizing the vast oil shale deposits of the Southwest. These deposits are almost entirely EROEI negative.

In an article entitled "A Crash Program for the Canadian Oil Sands Industry," in the journal *Energy Policy*, researchers Bengt Söderbergh, Fredrik Robelius and Kjell Aleklett meticulously analyze the production capabilities of the extensive Alberta Canada oil sands which are the planet's largest such deposit of this hydrocarbon resource. Their summary is clear:

> The implementation of a crash program for the Canadian oil sands industry is associated with serious difficulties. There is not a large enough supply of natural gas to support a future Canadian oil sands industry with today's dependence on natural gas. It is possible to use bitumen as fuel and for upgrading, although it seems to be incompatible with Canada's obligations under the Kyoto treaty. For practical long-term high production, Canada must construct nuclear facilities to generate energy for the in situ projects. Even in a very optimistic scenario Canada's oil sands will not prevent Peak Oil. A short-term crash program from the Canadian oil sands industry achieves about 3.6 mb/d by 2018. A long-term crash program results in a production of approximately 5 mb/d by 2030.[58]

The study is not very clear with respect to net EROEI for the oil produced from these deposits. It clearly acknowledges, however, that this figure will be low. Since natural gas will not be available in sufficient quantities to power the ever-increasing projected production from these deposits, the study concludes that the only possible option is to use on-site nuclear reactors for this purpose. Of course, nuclear power itself is problematic, as it will take at least a decade to build the reactors, and the reactors require an estimated 20 years operation to reach energy breakeven — the point at which the net energy produced by the reactor is equal to all of the energy required to construct it, mine and refine its nuclear fuel, and so on. Since the study makes it clear that reactor construction cannot be expected to start for at least a decade, it will be about 30 years before net energy would be produced from any on-site nuclear power sources.

But of course the reactor itself is only a power source to convert oil sand deposits into refined oil. This process is itself at best capable of producing only a very small net energy gain. Thus vast amounts of energy will have to be expended in the coming decades to produce a very small return on energy invested with the payback being decades into the future. This cannot be a meaningful solution to peak oil.

---

58. *A Crash Program Scenario for the Canadian Oil Sands Industry*, Bengt Söderbergh, Fredrik Robelius and Kjell Aleklett, Uppsala University, June 8, 2006, http://www.peakoil.net/uhdsg/20060608EPOSArticlePdf.pdf, pp 2.

The study envisions that under the most optimistic conditions imaginable, production of oil from these sands could reach 5 to 6 million barrels per day in about 2035 to 2040. After remaining at that level for an unspecified short time, apparently no more than a decade, production would begin a gradual but steady and inexorable decline. This means that just at the time that a net energy surplus is achieved, production begins to decline.

Also if we account for all of the energy required to produce this unconventional oil, and we assume, optimistically, that production actually can occur at a positive EROEI, then the net energy produced will not be 5-6 millions barrels of oil. After accounting for the energy required to produce this oil, the *net* energy production will be closer to about 1 million barrels per day. This won't even be enough for Canada's domestic needs!

The assertion that this resource provides the solution to peak oil — even with on-site nuclear power plants to "cook" the oil sands, is thus baseless. Under ideal conditions a small net amount of energy, in the form of liquid hydrocarbon fuels, can be produced for several decades from this resource. This will occur at the cost of vast environmental degradation and also at the cost of substantial CO_2 emissions which will, of course, further accelerate the process of global warming.

Additionally, given the small net energy payoff and the long timeframe required to achieve it, it is doubtful that a crash program to develop this resource is the best possible investment of scarce energy resources. Rather, by wasting these energy resources on this project, the process of crisis is likely to be intensified and other more productive uses of the diverted energy resources will not be implemented.

Yet this project certainly will be undertaken simply because it promises to produce vast wealth for energy corporations. Government subsidies and tax policies, which are determined by these corporations through their control over the political process in both Canada and the US — and elsewhere — will ensure that these corporations will always make a profit on every stage of this energy project. This profit will be further augmented by these corporations' complete control over pricing for the oil products produced by the project.

In summary, the study concludes that:

> Unfortunately, while the theoretical future oil supply from the oil sands is huge, the potential ability for the Canadian oil sands industry to meet expectations of bridging a future oil supply gap is not based on reality. Even if a Canadian crash program were immediately implemented it may only barely offset the combined declining conventional crude oil production in Canada and the North Sea. The more long-term oil sands production scenario outlined in this report, does not even manage to compensate for the decline by 2030....Finally it may be of interest to recapitulate that the International Energy Agency claims that 37 mb/d of unconventional oil must be produced by 2030. Canada has by far the largest unconventional oil reserves. By 2030, in a very optimistic scenario, Canada may produce 5 mb/d. Venezuela may perhaps achieve a production of 6 mb/d. Who will be the producers of

the remaining 26 mb/d? It is obvious that the forecast presented by the IEA has no basis in reality.[59]

The study's summary also discusses the planet's sole other major deposit of oil sands, which is located in Venezuela. The production capabilities for the South American oil sands are roughly similar with those of Canada's. The bottom line in EROEI terms is that oil sands, no matter how heavily they are promoted, are not — and cannot be — a solution to peak oil. At best, they can mitigate its effects slightly. At worst, they substantially exacerbate the entire set of peak oil related crises.

The technological and financial difficulty of exploiting these resources has not been fully appreciated by most energy companies and investors. An article dated July 8, 2006, on the British Times Online site states that:

> SHELL is facing a cost explosion in the expansion of the Athabasca Oil Sands Project, a mining venture that extracts oil from bitumen deposits in the Canadian province of Alberta. The first phase of expansion, intended to add 100,000 barrels daily to the current 155,000 barrel per day output was budgeted at C$7.3 billion (£3.6 billion) only a year ago. It is now expected to cost as much as C$11 billion, according to estimates published by Western Oil Sands, Shell's partner in the project. Shell Canada said yesterday that it was conducting an assurance review of the project's cost, pending a final investment decision later this year. Planned in three phases, the Athabasca expansion is intended to raise output to 500,000 bpd, and represents a large part of Shell's oil production ambitions. Shell admitted to "significant upward pressure on capital costs" but declined to confirm its partner's prediction of a 50 per cent increase.[60]

The same cost and EROEI limitations apply to creating synthetic oil from coal and natural gas using the immensely inefficient Fischer-Tropf coal gasification process,[61] although it will serve to keep the Strategy (1) war machine going for short while longer, just as was the case in World War II Germany. In fact, this process is so energy inefficient that it has only been implemented on an industrial scale on two occasions, by Nazi Germany during World War II and by South Africa during the 1980s to subvert crippling global sanctions on its apartheid regime. Industrial scale coal gasification also produces copious quantities of carbon dioxide, and therefore, it rapidly accelerates the process of global warming.

Nuclear power can produce substantial quantities of electrical energy; however, it is useless as a transportation fuel. What it can do is provide the energy to manufacture hydrogen which can indeed be used as a fuel. However, building nuclear power plants is very energy intensive, as is the mining and

59. Ibid., Söderbergh et. al., # 58, pp 41

60. *TimesOnline*, July 7, 2006, Costs explode at Shell Canadian venture, http://business.timesonline.co.uk/article/0,,13130-2259904,00.html

61. Gas and Oil.com, Arco, Syntroleum, and others develop pilot-scale GTL facility, http://www.gasandoil.com/goc/company/cnn74853.htm

refining of nuclear fuels. And of course these fuels exist in limited quantities themselves. Furthermore, careful analysis has demonstrated that nuclear power is limited by the availability of suitable ores. Below concentrations of about 200 grams of uranium per ton of rock, more energy is required to produce uranium fuel than is produced by this fuel. This limits any possible nuclear power option to providing no more than a small fraction of the power needs of the early 21st century for at most several decades.[62]

Nuclear energy is sometimes touted as being environmentally friendly since it does not directly produce greenhouse gases. This assertion is highly misleading. As one researcher explains:

> The advantage of nuclear power in producing lower carbon emissions holds true only as long as supplies of rich uranium last. When the leaner ores are used — that is, ores consisting of less than 0.01 percent (for soft rocks such as sandstone) and 0.02 percent (for hard rocks such as granite), so much energy is required by the milling process that the total quantity of fossil fuels needed for nuclear fission is greater than would be needed if those fuels were used directly to generate electricity. In other words, when it is forced to use ore of around this quality or worse, nuclear power begins to slip into a negative energy balance: more energy goes in than comes out, and more carbon dioxide is produced by nuclear power than by the fossil-fuel alternatives....There is enough usable uranium ore in the ground to sustain the present trivial rate of consumption — a mere 2 1/2 percent of all the world's final energy demand — and to fulfill its waste-management obligations, for around 45 years. However, to make a difference — to make a real contribution to postponing or mitigating the coming energy winter — nuclear energy would have to supply the energy needed for (say) the whole of the world's electricity supply. It could do so — but there are deep uncertainties as to how long this could be sustained. The best estimate (pretending for a moment that all the needed nuclear power stations could be built at the same time and without delay) is that the global demand for electricity could be supplied from nuclear power for about six years, with margins for error of about two years either way.[63]

Seriously addressing just the US's energy needs would require the rapid construction of hundreds of new nuclear plants. Given the economic and military crises which are clearly foreseeable during the second decade of the 21st century, it is improbable that such a Herculean effort will be made. If it is made, it is improbable that there will be sufficient time for it to be consummated. If it were to be consummated, supplies of nuclear fuel would soon run out.

Additionally, the mining and refining of the nuclear fuel in conjunction with the vast amount of construction required to build the nuclear power plants

---

62. Nuclear Power: The Energy Balance, Chapter 2 from ore to electricity, http://www.stormsmith.nl/Chap_2_Energy_Production_and_Fuel_costs_rev6.PDF
63. The Foundation for the Economics of Sustainability, April, 2006, Why Nuclear Power Cannot Be a Major Energy Source, http://www.feasta.org/documents/energy/nuclear_power.htm PDF version at: http://www.feasta.org/documents/energy/nuclear_power.pdf

would release enormous quantities of greenhouse gases — and other pollutants — into the atmosphere. If the entire planet were to embark upon a crash nuclear program, the result would be even more greenhouse gas emissions along with the total depletion of all positive EROEI nuclear fuel in about five years. In short, nuclear power does not, and cannot, solve the energy problems arising from peak oil.

These constraints also doom any possibility of a hydrogen economy. Hydrogen is the most abundant element in the universe. Unfortunately it does not exist in usable form on Earth; hydrogen is a form of stored energy. It must be manufactured, and this requires energy. Hydrocarbon energy could be used to manufacture hydrogen. However, this is foolish in EROEI terms since the manufacturing process would be a net loser. Three units of energy in the form of gasoline might be expended to manufacture two units of energy in the form of liquid hydrogen. The energy to manufacture it must come from nuclear energy if it is to be practical. However, as we've just seen, nuclear energy itself is not practical as a large-scale, long-term, energy source. The bottom line is that the so called hydrogen economy is a fantasy.

Recently, much has been made of ethanol and biodiesel as quick solutions to impending hydrocarbon liquids shortages. One item that is missing from most discussions of corn-derived ethanol as a hydrocarbon fuel replacement is that according to at least some researchers it possesses a negative EROEI:

> Neither increases in government subsidies to corn-based ethanol fuel nor hikes in the price of petroleum can overcome what one Cornell agricultural scientist calls a fundamental input-yield problem: it takes more energy to make ethanol from grain than the combustion of ethanol produces. At a time when ethanol-gasoline mixtures (gasohol) are touted as the American answer to fossil fuel shortages by corn producers, food processors and some lawmakers, Cornell's David Pimentel takes a longer range view. "Abusing our precious croplands to grow corn for an energy-inefficient process that yields low-grade automobile fuel amounts to unsustainable, subsidized food burning," said the Cornell professor in the College of Agriculture and Life Sciences. Pimentel, who chaired a US Department of Energy panel that investigated the energetics, economics and environmental aspects of ethanol production several years ago, subsequently conducted a detailed analysis of the corn-to-car fuel process.[64]

Additionally, Dr. Pimentel observes that growing fuel crops also requires using cropland which otherwise could be used to produce food. This topic is controversial, as other researchers have asserted that ethanol can be produced with a positive EROEI. Recent research by Pimentel and Tad W. Patzek has counter-argued, convincingly, that these claims are false. Summarizing their research, they state that:

---

64. Cornell Chronicle, Aug. 23, 2001, CU Scientist Terms Corn Based Ethanol 'Subsidized Food Burning', http://www.news.cornell.edu/Chronicle/01/8.23.01/Pimentel-ethanol.html

Energy outputs from ethanol produced using corn, switchgrass, and wood biomass were each less than the respective fossil energy inputs. The same was true for producing biodiesel using soybeans and sunflower, however, the energy cost for producing soybean biodiesel was only slightly negative compared with ethanol production. Findings in terms of energy outputs compared with the energy inputs were:

● Ethanol production using corn grain required 29% more fossil energy than the ethanol fuel produced.

● Ethanol production using switchgrass required 50% more fossil energy than the ethanol fuel produced.

● Ethanol production using wood biomass required 57% more fossil energy than the ethanol fuel produced.

● Biodiesel production using soybean required 27% more fossil energy than the biodiesel fuel produced (Note, the energy yield from soy oil per hectare is far lower than the ethanol yield from corn).

● Biodiesel production using sunflower required 118% more fossil energy than the biodiesel fuel produced.[65]

Their research provides a detailed refutation of contrary claims for a positive EROEI for ethanol and biodiesel fuels derived from the plants that could be used in the US.

Even if we assume that, somehow, the contrary claim — that the possibility of producing ethanol and or biodiesel with at least a slightly positive EROEI is valid, it would represent an inefficient way to produce fuel energy. Further, the deliberate loss of food-producing potential at a time when, as we shall see in subsequent chapters, famine is probable, are balanced against this hypothetical small EROI gain for these fuels, the case for ethanol and biodiesel weakens further.

Also it should be noted that ethanol has only 70 percent of the energy of a corresponding volume of gasoline. It requires about 1.42 gallons of ethanol to equal the amount of energy contained in a single gallon of gasoline. Direct comparisons of energy and price of the two fuels must take this into account, as a tank of ethanol is not the fuel equivalent of a tank of gasoline. It seems clear that, no matter what, neither ethanol nor biodiesel can substantially make up for declining oil and natural gas supplies.

Wind, solar, tidal energy, and other renewable forms of energy certainly have their place in a post-peak oil energy production network. However, they cannot substitute for declining oil and natural gas supplies.

---

65. Pimentel, Davis and Patzek, Yad W., Natural Resources Research, Vol. 14, N0.1, March 2005, Ethanol Production Using Corn, Switchgrass and Wood; Biodiesel Production Using Soybean and Sunflower, http://petroleum.berkeley.edu/papers/Biofuels/NRRe-thanol.2005.pdf

What can be done, and done quickly, is to burn much more coal. In addition to the declining EROEI involved in mining it, burning coal will release immense amounts of greenhouse gases into our planetary atmosphere and accelerate global warming. Thus it seems certain that one consequence of Strategy (2) above will be to rapidly accelerate the process of global warming.

The consequences of relying primarily upon coal for energy can be seen at work now in the People's Republic of China. Here, relatively large coal deposits are that nation's primary domestic hydrocarbon resource. As a consequence, China is burning ever larger quantities of domestic coal to fuel its rapidly growing economy.

> Unless China finds a way to clean up its coal plants and the thousands of factories that burn coal, pollution will soar both at home and abroad. The increase in global-warming gases from China's coal use will probably exceed that for all industrialized countries combined over the next 25 years, surpassing by five times the reduction in such emissions that the Kyoto Protocol seeks. The sulfur dioxide produced in coal combustion poses an immediate threat to the health of China's citizens, contributing to about 400,000 premature deaths a year. It also causes acid rain that poisons lakes, rivers, forests and crops.... Already, China uses more coal than the United States, the European Union and Japan combined. And it has increased coal consumption 14 percent in each of the past two years in the broadest industrialization ever. Every week to 10 days, another coal-fired power plant opens somewhere in China that is big enough to serve all the households in Dallas or San Diego. To make matters worse, India is right behind China in stepping up its construction of coal-fired power plants — and has a population expected to outstrip China's by 2030.[66]

Unfortunately, it will not just be China and India which rely heavily upon coal. The US possesses the world's largest coal reserves and will most certainly burn these reserves — or liquefy them — in a desperate attempt to maintain its economy amid ever worsening hydrocarbon liquid shortages. During desperate times things like pollution controls tend to fall by the way side. This means that as we come to rely more and more upon coal we will control its CO_2 emissions less and less.

Coal is being used to produce ethanol in industrial scale quantities in the US and abroad already:

> The town of Columbus, Nebraska, bills itself as a "City of Power and Progress." If Archer Daniels Midland gets its way, that power will be partially generated by coal, one of the dirtiest forms of energy. When burned, it emits carcinogenic pollutants and high levels of the greenhouse gases linked to global warming. Ironically this coal will be used to generate ethanol, a plant-based petroleum substitute that has been hyped by both environmentalists and President George Bush as the green fuel of the future. The agribusiness giant Archer Daniels Midland (ADM) is the largest US producer of ethanol, which it makes by distilling corn. ADM also operates coal-

---

66. *New York Times*, June 11, 2006, World Business: Pollution From Chinese Coal Casts a Global Shadow, http://www.nytimes.com/2006/06/11/business/worldbusiness/11chinacoal.html?_r=1&th&emc=th&oref=slogin

fired plants at its company base in Decatur, Illinois, and Cedar Rapids, Iowa, and is currently adding another coal-powered facility at its Clinton, Iowa ethanol plant. That's not all. "[Ethanol] plants themselves — not even the part producing the energy — produce a lot of air pollution," says Mike Ewall, director of the Energy Justice Network. "The EPA (US Environmental Protection Agency) has cracked down in recent years on a lot of Midwestern ethanol plants for excessive levels of carbon monoxide, methanol, toluene, and volatile organic compounds, some of which are known to cause cancer." A single ADM corn processing plant in Clinton, Iowa generated nearly 20,000 tons of pollutants including sulfur dioxide, nitrogen oxides, and volatile organic compounds in 2004, according to federal records. The EPA considers an ethanol plant as a "major source" of pollution if it produces more than 100 tons of any one pollutant per year, although it has recently proposed increasing that cap to 250 tons.[67]

Here we see how the coal industry and agribusiness come together to produce profits for both, due to the *cumulative* effects of government-subsidized high prices for ethanol[68] in conjunction with a host of separate and discrete government subsidies for coal production,[69] for agricultural production[70] and for ethanol production.[71]

The bottom line is that the taxpayers pay to enrich these corporations. In return, the corporations actually destroy net energy, while releasing vast quantities of pollutants which sicken or kill us, while they simultaneously accelerate the process of global warming. And, due to corporate control of the mass media, the taxpayers are conditioned to see all of this as environmental activism on the part of these selfsame energy and agribusiness corporations and to applaud it, while demanding more such "environmental" policies.

Burning coal deposits to produce energy will buy a small amount of time with respect to maintaining the energy foundation of national economies which possess coal reserves, such as those of China, India, and the US, but only at the cost of rapidly increasing atmospheric CO_2 levels in particular, and health-destroying pollution in general. As we shall see in subsequent chapters, this

67. *Alter Net*, June 7, 2006, The Dirty Truth About Green Fuel, http://www.alternet.org/envirohealth/37217/
68. Ethanol Mandates and Subsidies, http://www.pureenergysystems.com/news/2005/04/12/6900080_Acetone_and_Ester/Ethanol_Mandates_Subsidies.doc
69. Taxpayers For Common Sense, Fossil Fuel Subsidies: A Taxpayer Perspective, http://www.taxpayer.net/TCS/fuelsubfact.htm and hydrogen is a form of stored energy.
70. *Wikipedia, The Free Online Encyclopedia*: Agricultural Subsidies, http://en.wikipedia.org/wiki/Agricultural_subsidies; *The Columbia Encyclopedia*, 6th Edition 2001-2005, Agricultural Subsidies, http://www.bartleby.com/65/ag/agrisub.html; and Environmental Working Group's Farm Subsidy Data Base: United States, http://www.ewg.org:16080/farm/region.php?fips=00000.
71. M. King Hubbert Center for Petroleum Supply Studies, Energy and Dollar Costs of Ethanol Production With Corn, http://hubbert.mines.edu/news/Pimentel_98-2.pdf; and ZFacts.com, Know the Facts, Get the Source, Ethanol Production: at $7.24/Gas-Gallon-Saved?, http://zfacts.com/p/60.html.

creates another set of linked existential crises for the planet. And the coal reserves which appear to be immense at current extraction rates will quickly be depleted at much higher rates of extraction.

We have willingly bought into the assumption that nature is an infinite cornucopia which we can plunder without limit, and without any consequences to the natural systems within which human civilization is nested and upon which it is ultimately dependant for its very existence. We have chosen to become consumers rather than citizens.

The strategies which humanity's corporate-influenced political elites, most particularly those of the US, have chosen for dealing with the consequences of peak oil inexorably accelerates the process of economic decline and political disintegration. This occurs even as they simultaneously accelerate the ongoing process of global warming and environmental disruption.

The truth is that there is simply no way to maintain today's high energy consumption based political economy. There are no available substitutes for hydrocarbon energy. Further, we have delayed too long to successfully make an orderly transition to a lower net energy political economy. Worse, we are moving full speed ahead to expand our existing high energy consumption economy as though an infinite supply of hydrocarbon fuels existed.

These policy choices appear to be driving us to an inevitable and catastrophic collapse of civilization within the next several decades.

So long as our fundamental values remain unchanged, the progression of environmental catastrophe in tandem with political disintegration will not only continue, but will actually accelerate. This is because the global corporate-dominated political and economic system's materialistic values place instant gratification, wealth and comfort above all else, and are actually incompatible with physical reality.

We have bought into the belief that we humans are above and outside of nature. Our corporate-dominated civilizational complex adaptive memetic system is predicated upon such assumptions. However, these memes are not consistent with reality. They possess very low *fitness*. Therefore, our civilizational meme complex must either radically and rapidly transform itself into a fitter complex adaptive system, that is to say, it must evolve, or it shall inevitably face extinction. Either way, it is now simply too late to prevent enormous disruptions to our social, ecological and climatological environment from occurring as a consequence of our past actions. And as a direct consequence of these actions on our parts, there shall soon be all hell to pay. In a system you can never do only one thing.

Objective information about peak oil is available from the Association for the Study of Peak Oil and Gas (APSO), http://www.peakoil.net/. Additional links to numerous informative articles written by respected experts in their fields can be found at: http://www.oilcrisis.com.

CHAPTER 4. GLOBAL WARMING AND CLIMATE CHANGE: WIND AND WEATHER

Now there is one outstandingly important fact regarding Spaceship Earth, and that is that no instruction book came with it. — R. Buckminster Fuller[72]

UNIQUE EARTH

The earth offers a specific set of conditions that fostered the evolution of life as we know it: an abundance of carbon and oxygen (and a limited amount of more noxious substances), water, a relatively stable climate for the last 10,000 years, and so forth. It took a long time to form, but it won't take long to destroy.

Eukaryotic cells[73] evolved perhaps a billion and a half years ago. Eukaryotic cells possess a definite membrane-bound cellular nucleus. They are perhaps a thousand times larger than are their prokaryotic ancestors. They feature complex internal structure including mitochondria which "power" the cell and organelles which perform specialized functions. This form of cell uses oxygen for its metabolism. This form of metabolism is vastly more energetic than the simple metabolic pathways utilized by prokaryotic organisms. Biology's taxonomic kingdoms of animals, plants, fungi (usually multi-cellular), and protists (usually unicellular) are composed of eukaryotic organisms.

Oxygen was produced in the earth's very early biosphere by photosynthesizing bacteria; however, billions of years were required for it to build up in the earth's atmosphere because the earth's crust absorbed it as it was produced.

---

72. Wisdom Quotes.com: Buckminster Fuller, http://www.wisdomquotes.com/002658.html
73. *Thinkquest.org*, Cellupedia, Eukaryotic Cells, http://library.thinkquest.org/C004535/eukaryotic_cells.html

About 2.3 billion years ago the crust became saturated with oxygen, finally allowing it to begin to accumulate in the atmosphere.[74][75] This buildup was gradual during the next billion and an half or so years. Once significant quantities of oxygen existed in the atmosphere, and was dissolved into sea water, a new metabolic pathway for exploiting it became possible. This development facilitated the evolution of eukaryotic cells. Evidence suggests that the accumulation of significant quantities of oxygen in sea water was extraordinarily slow:

> New measurements by University of Rochester geochemists have uncovered evidence that even after 2.2 billion years ago, the amount of oxygen in the oceans remained low, perhaps up to the time when multicelled life began to proliferate a few hundred million years ago....[This research] suggests the possibility that the relatively recent in rise in oxygen in the ocean might have been an important environmental stimulus for the evolution of multicelled life.... A world full of anoxic [low oxygen content] oceans could have serious consequences for evolution. Eukaryotes, the kind of cells that make up all organisms except bacteria, appear in the geologic record as early as 2.7 billion years ago, but multicelled eukaryotes did not appear until much later. One of the paper's authors, Ariel Anbar of the University of Rochester, previously suggested (with paleontologist Andrew Knoll of Harvard University) that an extended period of anoxic oceans might be the key to why the more complex eukaryotes barely eked out a living while their prolific bacterial cousins thrived.[76]

Complex multi-cellular life suddenly appeared and radiated into a vast profusion of forms in the Cambrian explosion which occurred between 542 and 530 million years ago.[77] Available evidence strongly suggests that this event was triggered by the concentration of dissolved oxygen in seawater suddenly reaching a critical threshold beyond which multi- cellular eukaryotic organisms could exploit it successfully for their metabolism. The era of animals and plants was inaugurated.

Several hundred million years passed before life took hold on land. Possibly the atmosphere's oxygen content had to build up further to allow this to occur. Overall, more than half a billion years would pass before a complex self-conscious organism able to create a civilization that could exploit hydrocarbon energy — humanity — evolved.

---

74. *Science Now*, 3 Aug, 2001, The Tardy Origin of Earth's Atmosphere, http://bric.postech.ac.kr/science/97now/01_8now/010803c.html

75. *Science* 3 August 2001: Vol. 293. no. 5531, pp. 839 – 843, Biogenic Methane, Hydrogen Escape, and the Irreversible Oxidation of Early Earth, David C. Catling, Kevin J. Zahnle, Christopher P. McKay.

76. National Aeronautics and Space Administration, NASA Astrobiology Institute, March 8, 2004, When Did Earth's Oceans Become Oxygenated? http://nai.arc.nasa.gov/news_stories/news_detail.cfm?ID=280

77. *Encyclopædia Britannica*, 2006, Community ecology, Encyclopædia Britannica Premium Service, http://www.britannica.com/eb/article-70649.

Billions of years of warmth that had allowed liquid water to continue to exist without completely freezing solid or boiling away, low cosmic background radiation, low frequency of bombardment of the planet by asteroids and comets, solar stability, and many other variables being just right, were needed to allow for the evolution of an intelligent, potentially space-faring species on Earth.

If the sun were larger it would have gone nova by now — not allowing enough time for a species such as humanity to develop. If it were smaller and hence dimmer, the habitable zone in its orbit, where water remains liquid, would be much smaller. Thus, the chances that a planet would form at just the right distance from the sun within this narrow habitable zone would be much reduced. An additional problem occurs because all stars heat as they age, thus the time that any planet could occupy this habitable zone would be far less than the billions of years needed for intelligent life to evolve. Additionally, Earth is fortunate to have an orbit around the galactic center which keeps it away from dangerous areas of radiation:

> The star must be in a galactic orbit with a period that generally keeps it out of the galaxy's spiral arms, which have more frequent supernova, which create radiation hazards. If the orbit is eccentric (egg-shaped), it will pass through the arms. The star must also orbit well away from the core. An orbit that takes it too near an energetic galactic core will expose it to hard radiation. The Sun is in a nearly perfect circular orbit with a period of 226 million years, in a narrow ring of orbits whose periods nearly match the rotational period of the galaxy. The Sun's galactic orbit is so perfect that it has remained outside the galactic arms for more than 18 orbits. Our star has to be in the suburbs of the galaxy; it cannot be in the city or the countryside.[78]

Earth's galactic orbit is optimal for life — however this is not true for the vast majority of stars in this or any other galaxy.

Additionally, a large moon to keep the planet stable on its axis of rotation is needed. Otherwise the planet would continuously precess on its plane of rotation randomly from zero to ninety degrees. The resulting climatic chaos would prevent complex life forms from ever getting established because the climate would never be stable. Equatorial regions would become polar ones and so on, endlessly. Additionally, vigorous plate tectonics is also needed for complex life to evolve. Ward and Brownlee offer four reasons as to who plate tectonics is so critically important:

> First, plate tectonics promotes high levels of global biodiversity. In the last chapter we suggested that major defense against mass extinctions is high biodiversity. Here we argue that the factor on Earth that is most critical to maintaining diversity through time is plate tectonics. Second, plate tectonics provides our planet's global thermostat by recycling chemicals crucial to keeping the volume of carbon dioxide in our atmosphere relatively uniform, and thus it has been the single most impor-

---

78. Space.com, Rare Earth Debate: Part 1: The Hostile Universe, 15 July, 2002, http://www.space.com/scienceastronomy/rare_earth_1_020715.html

tant mechanism enabling liquid water to remain on Earth's surface for more than 4 billion years. Third, plate tectonics is the dominant force that causes changes in sea level, which, it turns out, are vital to the formation of minerals that keep the global level of carbon dioxide (and hence global temperature) in check. Fourth, plate tectonics created the continents on planet earth. Without plate tectonics Earth might look much as it did during the first billion and a half years of its existence: a watery world, with only isolated volcanic islands dotting its surface. Or, it might look even more inimical to life; without continents, we might by now have lost the most important ingredient for life, water, and in so doing come to resemble Venus. Finally plate tectonics makes possible one of Earth's most potent defense systems: its magnetic field. Without our magnetic field Earth and its cargo of life would be bombarded by a potentially lethal influx of cosmic radiation, and solar wind "sputtering" (in which particles from the sun hit the upper atmosphere with high energy) might slowly eat away the atmosphere, as it has on Mars.[79]

It appears that the evolution of sentient life is an astronomically rare event. Furthermore, although the sun has about five billion years left before it exhausts its hydrogen fuel, goes nova, incinerates the earth, and becomes a white dwarf, life on Earth — at least complex life — has much less time left. As the sun ages, it grows progressively hotter. During the lifetime of the earth the sun's heat has increased by about 30%. In the meantime the amount of heat-trapping carbon dioxide in the atmosphere has gradually declined.

As can be seen from Figure 1, most research has found that carbon dioxide levels have gradually decreased across hundreds of millions of years, while solar intensity has increased. The net effect has been to maintain temperatures on the surface of the earth which allow for water to remain liquid. However, as the sun continues to heat up, the capability of the earth's biosphere to facilitate further carbon dioxide reductions to continue to keep the planet within habitable temperatures for life is coming to an end as CO_2 levels approach zero. Within 500 million to, at the very most, one billion years from now, surface temperature will therefore rise to above the boiling point of water and complex life on the planet will become extinct, though microbes deep below the surface will most likely continue to exist until the sun goes nova. Figure 2 below illustrates this.

The relatively stable climate of the past 10,000 years has allowed for agriculture and thus civilization to take hold. This stable climate is an aberration in the recent history of earth. Such a period may well not come again for geological ages.

79. Ward, Peter D., and Brownlee, Donald, *Rare Earth: Why Complex Life is Uncommon in the Universe*, Springer-Verlag, New York, NY, 2000, pp, 194.

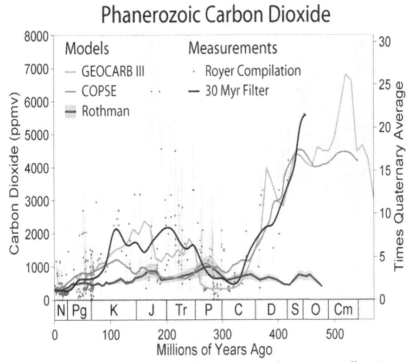

Figure 1. Atmospheric Carbon Dioxide Content Over the Last 500 Million Years.

Source: Wikipedia, The Free Online Encyclopedia.[80]

GLOBAL OVERVIEW

The Earth's biosphere, the area within which life exists, extends from about six miles into the air to the ocean bottom — more than six miles below sea level. Water is inherently a much denser medium than air. Compare a cubic yard (or meter) of air with a corresponding volume of water. There are about one hundred times as many atoms in the volume of water than in the corresponding volume of air.

---

80. *Wikipedia, The Free Online Encyclopedia:* Image: Phanerozoic Carbon Dioxide, Permission is granted to copy, distribute and/or modify this document under the terms of the GNU Free Documentation License, Version 1.2 or any later version published by the Free Software Foundation; with no Invariant Sections, no Front-Cover Texts, and no Back-Cover Texts, http://en.wikipedia.org/wiki/Image:Phanerozoic_Carbon_Dioxide.png

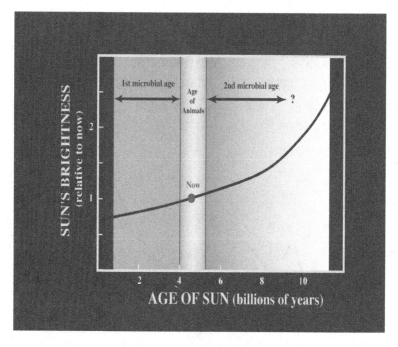

Figure 2. Duration of Complex Life on Earth.

Source: Peter Ward and Don Brownlee, University of Washington[81][82]

In terms of physical measurements, the oceans comprise a volume of about 1.37 billion cubic kilometers, or 328 million cubic miles.[83] The Earth's atmosphere lacks a definite boundary, however, at 11 kilometers, about the height a jet cruises at (between six and seven miles altitude) 77.5 percent of the atmosphere lies below. We can consider this to be the upper limit of the biosphere. The total mass of the atmosphere is about 5.3 times ten to the eighteenth power kilograms.[84] This is obviously a large number; however it is less than one one-hundredth of the mass of the oceans. While there is a clear demarcation between

81. Ward, Peter D., and Brownlee, Donald, *The Ends of the Earth Astrobiology & Armageddon*, Peter Ward and Don Brownlee, University of Washington, http://www.astro.washington.edu/endsofworld/

82. Ward, Peter D., and Brownlee, Donald, *The Life and Death of Planet Earth, How The New Science of Astrobiology Charts The Ultimate Fate of Our World*, Henry Holt and Company LLC, New York, NY, 2002, pp. 106.

83. Ellert, Glen, Ed., *The Physics Factbook*, Volume of the Earth's Oceans, http://hypertextbook.com/facts/2001/SyedQadri.shtml

84. Public Domain Aeronautical Software: How Thick is the Earth's Atmosphere?, http://www.pdas.com/atmthick.htm

atmosphere and ocean, the atmosphere also contains evaporated water — water vapor. This comprises anywhere between zero to four percent of the atmosphere, by volume.[85]

Weather, including oceanic phenomena such as deep ocean currents, is caused by temperature differentials within and between the atmosphere and the ocean. If the entire surface of the earth consisted of water of a uniform depth which was uniformly heated across its entire area, no difference in temperature would exist. In accordance with the laws of thermodynamics, energy flows from higher potential to lower potential. If there were no temperature differences, there would be no potential differences, and thus there would be no weather!

Uneven heating of the earth's atmosphere, oceans, and other bodies of water, are the cause of all weather. The planet's weather is thus a means for temperature equalization across unequally heated mediums of air and water. Such heat redistribution occurs chaotically.

> [C]haos theory deals with the behavior of certain nonlinear dynamic systems that under certain conditions exhibit a phenomenon known as chaos, which is characterized by a sensitivity to initial conditions (see butterfly effect). As a result of this sensitivity, the behavior of systems that exhibit chaos appears to be random, even though the model of the system is deterministic in the sense that it is well defined and contains no random parameters.[86]

Chaotic systems are inherently sensitive to initial conditions. This sensitivity is popularly known as the "butterfly effect." Practically, this means that any arbitrarily slight change in conditions will cause totally divergent outcomes within a short time:

> The butterfly effect is a phrase that encapsulates the more technical notion of sensitive dependence on initial conditions in chaos theory. The idea is that small variations in the initial conditions of a dynamical system produce large variations in the long term behavior of the system. Sensitive dependence is also found in non-dynamical systems: for example, a ball placed at the crest of a hill might roll into any of several valleys depending on slight differences in initial position. The practical consequence of the butterfly effect is that complex systems such as the weather are difficult to predict past a certain time range — approximately a week, in the case of weather. This is because any finite model that attempts to simulate a system must necessarily truncate some information about the initial conditions — for example, when simulating the weather, one would not be able to include the wind coming from every butterfly's wings. In all practical cases, defects in the knowledge of the initial conditions and deficiencies in the model are equally important sources of error. In a chaotic system, these errors are magnified as the simulation progresses. Thus the predictions of the simulation are useless after a certain finite amount of time. Edward Lorenz first analyzed the effect in a 1963 paper for the New York

85. Royal BC Museum, Composition and Structure of the Atmosphere, http://www.livinglandscapes.bc.ca/thomp-ok/env-changes/atmos/ch2.html

86. *Encyclopædia Britannica*, 2006, Analysis, Encyclopædia Britannica Premium Service, http:/www.britannica.com/eb/article-218282.

Academy of Sciences. According to the paper, "One meteorologist remarked that if the theory were correct, one flap of a seagull's wings could change the course of weather forever."[87]

This means that while the planet's weather patterns do obey deterministic physical laws, for all intents and purposes they are not predictable far in advance. It also means that even fairly small changes in long-established heating or convection patterns can suffice to dramatically alter all subsequent weather patterns, globally and permanently.

The biosphere — the breathable atmosphere, oceans and other bodies of water, along with the land surface, represents a finely tuned set of interactions between living and non-living variables: the oxygen in the air, along with trace gases such as carbon dioxide and methane, are products of living systems.

Because oxygen is chemically very reactive, meaning it is very prone to bind chemically with other substances, it would quickly be removed from the planet's atmosphere unless it was continually replenished by living things. Atmospheric oxygen is entirely a product of life.

Atmospheric moisture determines rainfall on land, and thus the habitability of particular areas of the land. Yet, the presence of plants on land alters the area's albedo — its reflectivity to sunlight. This in turn determines its degree of temperature differential with surrounding areas, and hence its probability of receiving moisture laden atmospheric currents.

Energy reaches the surface of the earth, heating it, primarily in the visible spectrum because our planet's atmosphere is transparent to electromagnetic radiation at those particular wavelengths while being opaque to most other electromagnetic wavelengths. Figure 1, just below shows this.

This, incidentally, accounts for why our eyes are sensitive to only that small portion of the vast electromagnetic spectrum that we refer to as comprising visible light.

There are several ways in which humans can affect the planet's temperature:

1) Emission of greenhouse gases. These gases are transparent to incoming light at or near the visible spectrum. However, some of this incoming energy is subsequently reradiated back into space as infrared radiation — heat. Greenhouse gases act by allowing the energy in as light while blocking it from leaving again in the form of infrared or heat, radiation.

2) Changing the reflectivity (albedo) of the planet's surface. Snow and ice cover large portions of the polar regions of the planet. They act somewhat like mirrors in that they reflect much of the light energy which reaches them back out into space. Note that since the light is not absorbed first, it is not transformed into infrared, heat, energy. It remains visible spectrum energy. As such, it

---

87. *Encyclopædia Britannica*, 2006, Complexity, Encyclopædia Britannica Premium Service, http://www.britannica.com/eb/article-129415.

is efficiently reradiated back into space regardless of the presence of greenhouse gases. This occurs when, for example, increasing quantities of greenhouse gases accumulate in the atmosphere, causing the surface temperature to rise. This rise is most pronounced at the poles — where most of the very reflective ice and snow is also located. Increasing temperatures cause some of this ice and snow to melt. This exposes the bare ground underneath which is much darker, and hence absorbs the light which strikes it rather than reflecting it back into space. This absorption causes further warming which causes still more ice and snow to melt triggering yet more warming. And so on.

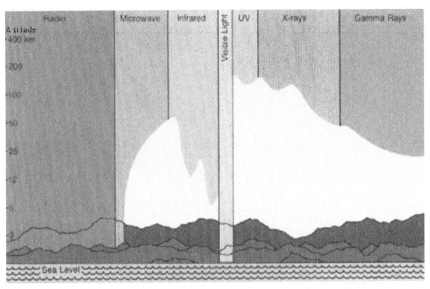

Figure 3. Electromagnetic Absorption Spectrum of Atmosphere.
Source: NASA.[88]

3) Emitting particulate matter into the atmosphere. Smoke, soot, smog, other chemical hazes, all work to block visible light from ever reaching the surface of the earth. The effect of this is to reduce temperatures since less energy reaches the earth's surface. Conversely, a reduction in atmospheric particulate matter acts to raise surface temperatures.

Another critical process which we need to understand in order to properly evaluate the effects of human activities upon the planet's natural cycles is called the carbon cycle. Figure 2, just below illustrates it.

---

88. NASA, Electromagnetic Spectrum, http://imagine.gsfc.nasa.gov/docs/science/know_ll/emspectrum.html

Note that the numbers in the figure just above refer to gigatons (billions of tons) of carbon. Carbon is essential to life on Earth. Its atomic structure, characterized by four unbound (valence) electrons, makes it better able to form a larger array of chemical arrangements with other atoms than any other atom. Life requires very complex chemistry. Carbon is the irreplaceable core of this complexity. Carbon cycles throughout the biosphere as shown in figure two. Human actions which interfere with this cycling often cause additional carbon to be released into the atmosphere, which otherwise would have remained sequestered within the cycle. Carbon in the atmosphere takes the form of carbon dioxide — a potent greenhouse gas.

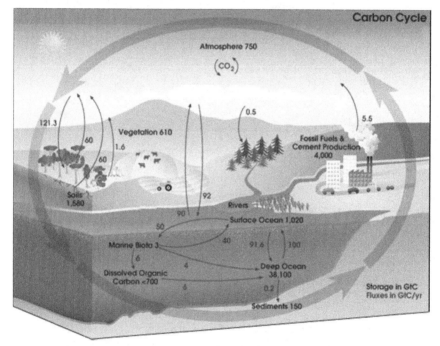

Figure 4. The Carbon Cycle.

Source: NASA.[89]

With our understanding of these basic facts, we are now ready to investigate the effects of humans upon nature. Nature is a complex adaptive system, as we've discussed above; it is of immense complexity, characterized by countless feedback and feedforward loops, such as the carbon cycle. Nature,

---

89. NASA, Earth Observatory, The Carbon Cycle, The Human Influence, http://earthobservatory.nasa.gov/Library/CarbonCycle/carbon_cycle4.html

which is the totality of these constantly interacting cycles, embodies a dynamic balance which has maintained the conditions necessary for life — including human life — to flourish for geological ages. However, it is sensitive to initial conditions, as we've already noted.

Civilization is now acting to alter many of those initial conditions in a very dramatic manner. In 2004 an unpublished Pentagon report on the consequences of global warming over the next several decades concluded that: "Climate change over the next 20 years could result in a global catastrophe costing millions of lives in wars and natural *disasters*..."[90] According to the *London Observer*,

> A secret report, suppressed by US defence chiefs and obtained by The Observer, warns that major European cities will be sunk beneath rising seas as Britain is plunged into a "Siberian" climate by 2020. Nuclear conflict, mega-droughts, famine and widespread rioting will erupt across the world. The document predicts that abrupt climate change could bring the planet to the edge of anarchy as countries develop a nuclear threat to defend and secure dwindling food, water and energy supplies. The threat to global stability vastly eclipses that of terrorism, say the few experts privy to its contents. "Disruption and conflict will be endemic features of life," concludes the Pentagon analysis. "Once again, warfare would define human life."[91]

SOWING THE WIND

In a system any change anywhere ripples through it affecting all parts of the system. Perhaps the direst consequence of peak oil is rapid global climate change. Consider the following question and its answer, which appeared in a recent edition of *Newsweek*:

MARGOLIS: Are natural disasters getting worse?

BOGARDI: There are absolutely clear signs and compelling statistics showing the situation is getting worse. We now are experiencing 2.5 to 3 times as many extreme events of climatic or water-related emergencies per year as we did in the 1970s. At the same time annual economic losses [from disasters] have increased sixfold.[92]

Rapid burning of hydrocarbons releases increasing amounts of carbon dioxide (CO_2) and other greenhouse gases into our atmosphere. These trap solar radiation and consequentially, the earth warms. Figures 5 and 6 below illustrate this relationship:

---

90. *The Observer*, Feb. 22, 2004, Now the Pentagon tells Bush: Climate change will destroy us, http://observer.guardian.co.uk/international/story/0,6903,1153513,00.html
91. Ibid., *The Observer*, # 90
92. *Newsweek International Edition*, Oct, 31, 2005, The Last Word: Preparing for the Worst, http://www.msnbc.msn.com/id/9785606/site/newsweek/

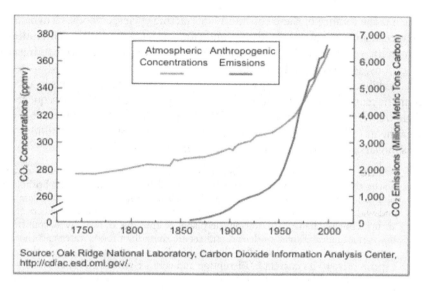

Figure 5. Atmospheric Carbon Dioxide Concentrations: 1750-2000.

Source: Oak Ridge National Laboratory[93]

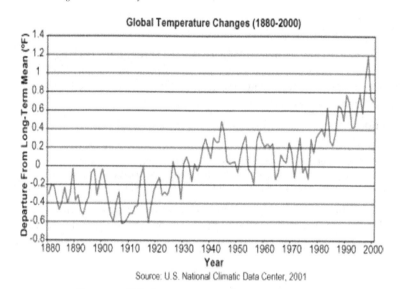

Figure 6. Global Temperature Changes: 1880-2000.

Source: US Environmental Protection Agency[94]

---

93. Energy Information Administration, Greenhouse Gases, Climate Change, and Energy, http://www.eia.doe.gov/oiaf/1605/ggccebro/chapter1.html

The projected future temperature increases caused by projected future increases in CO_2 are shown in Figure 5 just below:

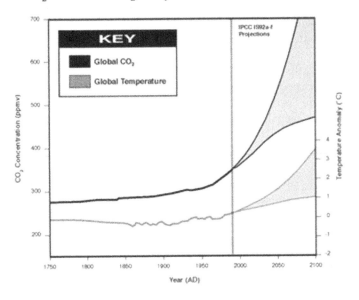

Figure 7. Projected CO_2 Caused World Temperature Changes: 1990-2100.

Source: US Global Change Climate Research Program[95]

The basic relationship seems simple enough. More CO_2 in the atmosphere means that more heat is trapped in the earth's atmosphere, resulting in higher mean global temperatures. A glance at any newspaper in December or January of each year confirms this warming trend annually:

> This year has been one of the hottest on record, scientists in the United States and Britain reported yesterday, a finding that puts eight of the past 10 years at the top of the charts in terms of high temperatures.[96]

> Three studies released yesterday differ slightly, but they all indicate the earth is rapidly warming. NASA's Goddard Institute for Space Studies has concluded 2005 was the warmest year in recorded history, while the National Oceanic and Atmospheric Administration and the U.K. Meteorological Office call it the second hottest, after

94. US Environmental Protection Agency, Global Warming-Climate, http://yosemite.epa.gov/oar/globalwarming.nsf/content/Climate.html

95. US Global Change Research Program, Anticipated Climate Changes in a High CO_2 World: Implications for Long-Term Planning, USGCRP Seminar, 15 September, 1997, http://www.usgcrp.gov/usgcrp/seminars/9799DD.html

96. *Washington Post.com*, Dec. 16, 2005, 2005 Continues the warming trend, year's temperatures are among the highest on record, scientists announce, http://www.washingtonpost.com/wp-dyn/content/article/2005/12/15/AR2005121501637.html

1998. All three groups agree that 2005 is the hottest year on record for the Northern Hemisphere, at roughly 1.3 degrees Fahrenheit above the historical average.[97]

Things are actually considerably worse than this simplistic model suggests, because there are other human-caused greenhouse gases besides CO_2, such as methane and nitrous oxides, which also contribute towards overall warming.

Worse still, anthropogenic or human-caused warming is now triggering the rapid release of vast quantities of methane and other greenhouse gases, which certainly produce even more rapid warming. The potential exists for runaway warming to be set in motion. When that happens, then global warming will have become both irreversible and will have moved completely beyond human control:

> Climate scientists yesterday reacted with alarm to the finding, and warned that predictions of future global temperatures would have to be revised upwards. "When you start messing around with these natural systems, you can end up in situations where it's unstoppable. There are no brakes you can apply," said David Viner, a senior scientist at the Climatic Research Unit at the University of East Anglia. "This is a big deal because you can't put the permafrost back once it's gone. The causal effect is human activity and it will ramp up temperatures even more than our emissions are doing." In its last major report in 2001, the intergovernmental panel on climate change predicted a rise in global temperatures of 1.4C-5.8C between 1990 and 2100, but the estimate only takes account of global warming driven by known greenhouse gas emissions. "These positive feedbacks with land-masses weren't known about then. They had no idea how much they would add to global warming," said Dr Viner. Western Siberia is heating up faster than anywhere else in the world, having experienced a rise of some 3C in the past 40 years. Scientists are particularly concerned about the permafrost, because as it thaws, it reveals bare ground which warms up more quickly than ice and snow, and so accelerates the rate at which the permafrost thaws.[98]

Recall that not all feedback is linear. Under certain conditions, a small change in input can produce a very large change in output. This is how amplification works. Recall also that complex adaptive systems such as the planet's weather and the biosphere as a whole are characterized by sensitive dependence on initial conditions. This nonlinear response to the inputs is taking place right now with the biosphere. Still, things are even worse:

> A study of 96 watersheds in western Siberia is showing an alarming trend: If temperatures in the region and throughout the Arctic continue to rise, by the end of this century, land surface that is covered by permafrost will be halved — leaching carbon into the area's water system. Considering that vegetation throughout West Siberia contains one-third of the carbon buried in the world's soils, such a perma-

---

97. Real Climate: Climate science from climate scientists, 15 Dec. 2005, 2005 temperatures, http://www.realclimate.org/index.php?p=231

98. *The Observer*, Aug. 11, 2005, Warming hits 'tipping point', http://www.guardian.co.uk/climatechange/story/0,12374,1546824,00.html

frost loss could dramatically affect the global climate cycle. Most studies of the effects of melting Arctic permafrost tend to focus on direct emissions to the atmosphere of carbon dioxide and methane from the peat — compacted undecomposed plant matter — newly exposed beneath the melting permafrost. But, says Karen Frey, a hydrologist at the University of California, Los Angeles (UCLA), streams and rivers could also lead to atmospheric carbon dioxide emissions by carrying once-buried carbon out of the peat and into the ocean....Dissolved organic carbon leaches out of peat and is dumped into the Arctic Ocean, where bacteria decompose the carbon material over a decade. The carbon is then released as carbon dioxide into the atmosphere, thus providing yet another positive feedback system: As Earth warms, dissolving and decomposing peat releases more carbon into the atmosphere, thus warming Earth more, says Robert Holmes of the Woods Hole Research Center in Massachusetts.[99]

Human-triggered rapid climate change unleashes still more of nature's positive (amplifying) feedback loops. Warming increases yet more rapidly. Worse, the cascading effects of greenhouse gases being released by melting at some point become self-sustaining. From then onwards, nature behaves like a runaway train. Humans have no control over further warming.

By the beginning of 2006 evidence that this was occurring was mounting alarmingly:

> Global warming is set to accelerate alarmingly because of a sharp jump in the amount of carbon dioxide in the atmosphere. Preliminary figures, exclusively obtained by The Independent on Sunday, show that levels of the gas — the main cause of climate change — have risen abruptly in the past four years. Scientists fear that warming is entering a new phase, and may accelerate further....Through most of the past half-century, levels of the gas rose by an average of 1.3 parts per million a year; in the late 1990s, this figure rose to 1.6 ppm, and again to 2ppm in 2002 and 2003. But unpublished figures for the first 10 months of this year show a rise of 2.2ppm. Scientists believe this may be the first evidence that climate change is starting to produce itself, as rising temperatures so alter natural systems that the earth itself releases more gas, driving the thermometer ever higher.[100]

Atmospheric CO_2 content has begun rising at a rate that gets ever faster.

The rate of increase is accelerating from 1.3 ppm as recently as a decade ago to 2.2 ppm in 2005; with the rate of increase itself increasing ever more rapidly. Greenhouse gas release is now entering a self-sustaining positive feedback phase. All of our assumptions, all of our models based upon a gradual, linear, rises in greenhouse gases, are now out the window.

In March of 2006 US and British climate scientists announced that their analysis of worldwide atmospheric data for 2005 showed that atmospheric CO_2 content had risen during 2005 to 381 ppm. This represented an increase of 2.6

99. *Geotimes*, July 2005, News Notes, Hydrology, Carbon leaching out of Siberian peat, http://www.geotimes.org/july05/NN_arcticpeatCO2.html

100. The Independent, 15 Jan. 2006, Global warming to speed up as carbon levels show sharp rise, http://news.independent.co.uk/environment/article338689.ece

ppm for that year. This further documented the rapid rise in the rate at which atmospheric CO_2 was increasing.

> "We don't see any sign of a decrease; in fact, we're seeing the opposite, the rate of increase is accelerating," Dr Pieter Tans told the BBC. The precise level of carbon dioxide in the atmosphere is of global concern because climate scientists fear certain thresholds may be "tipping points" that trigger sudden changes. The UK government's chief scientific adviser, Professor Sir David King, said the new data highlighted the importance of taking urgent action to limit carbon emissions. "Today we're over 380 ppm," he said. "That's higher than we've been for over a million years, possibly 30 million years. Mankind is changing the climate."[101]

About one week later the World Resources Institute published a major study of climate change entitled "Climate Science 2005 Major New Discoveries." The study analyzed all of the major climate research and discoveries for 2005. Their conclusions were as unambiguous as they were stark:

> The findings reported in peer-reviewed journals last year [2005] point to an unavoidable conclusion: The physical consequences of climate change are no longer theoretical; they are real, they are here, and they can be quantified...Taken collectively [these findings] suggest that the world may well have moved past a key physical tipping point....the new scientific findings reviewed here (coupled with the overall trend of rapid increases in greenhouse gas emissions) are any indicator, they suggest the world is in for both an ominous report [the 2006 Intergovernmental Panel on Climate Change Report], and more significantly, a major shift in Earth's climate....even if our society were to halt greenhouse gas emissions today, we have already committed to substantial warming and sea level rise in future years.[102]

These findings are clear and unambiguous. Particularly compelling is the conclusion that global warming and its associated climate change are now unstoppable. A study published in the 18 March 2005 edition of *Science* entitled "The Climate Change Commitment" demonstrated that thermal inertia in the oceans alone, absent any continuing anthropogenic greenhouse gas production, could raise mean global temperatures by between 2 and 6 degrees Celsius, or 3.6 to 9.8 degrees Fahrenheit by the year 2400.[103]

John D. Cox, a veteran science journalist writes in his recent book *Climate Crash* that:

> [T]he climate system is non-linear, which means its output is not always proportional to its input — that occasionally, unexpectedly, tiny changes in initial conditions provoke huge responses. It is chockablock with feedback loops, loops of self-perpetuating physical transactions, operating on their own timescales that amplify

---

101. BBC News International, 14 March, 2006, Sharp Rise in CO_2 Levels Recorded, http://news.bbc.co.uk/2/hi/science/nature/4803460.stm

102. World Resources Institute, Climate Science 2005, Major New Discoveries, March 2006, http://pdf.wri.org/climatescience_2005.pdf

103. The Climate Change Commitment, *Science*, 18 March, 2005, 307(5716): 1766-1769, http://www.sciencemag.org

or impede other processes. This constant crosstalk of positive and negative feedbacks is said to be balanced, more or less, at various critical thresholds in the system. Forced across such a threshold, by whatever external or internal triggering mechanism, important variables begin gyrating or flickering, and the system suddenly lurches into a significantly different semi-stable mode of operation, and a new equilibrium.[104]

At the same time that we are rapidly pushing nature out of its long-term equilibrium, the willingness of the corporate-dominated governments of major nations to deal with these human caused problems, even upon being presented with irrefutable evidence that the process of global warming and rapid climate change is now spiraling out of control, is effectively nonexistent:

> But a summit of the most polluting countries, convened by the Bush administration, last week refused to set targets for reducing their carbon dioxide emissions. Set up in competition to the Kyoto Protocol, the summit, held in Sydney and attended by Australia, China, India, Japan and South Korea as well as the United States, instead pledged to develop cleaner technologies — which some experts believe will not arrive in time.[105]

James Lovelock, originator of the Gaia hypothesis, hypothesized that the earth possesses a planetary control mechanism which has kept it habitable for geological ages. By early 2006, this renowned climate scientist had concluded that it is now too late to prevent sudden and rapid climate change set in motion by human caused actions:

> The world has already passed the point of no return for climate change, and civilisation as we know it is now unlikely to survive, according to James Lovelock, the scientist and green guru who conceived the idea of Gaia — the earth which keeps itself fit for life. In a profoundly pessimistic new assessment, published in today's Independent, Professor Lovelock suggests that efforts to counter global warming cannot succeed, and that, in effect, it is already too late. The world and human society face disaster to a worse extent, and on a faster timescale, than almost anybody realises, he believes. He writes: " Before this century is over, billions of us will die, and the few breeding pairs of people that survive will be in the Arctic where the climate remains tolerable"....And in today's Independent he writes: "We will do our best to survive, but sadly I cannot see the United States or the emerging economies of China and India cutting back in time, and they are the main source of [CO_2] emissions. The worst will happen ..." He goes on: "We have to keep in mind the awesome pace of change and realise how little time is left to act, and then each community and nation must find the best use of the resources they have to sustain civilisation for as long as they can." He believes that the world's governments should plan to secure energy and food supplies in the global hothouse, and defences against the expected rise in sea levels. The scientist's vision of what human society may ulti-

104. Cox, John D., *Climate Crash: Abrupt Climate Change and What it Means for our Future*, Joseph Henry Press, pp. 147, Washington, D.C., 2005

105. *The Independent*, Jan. 15, 2006, Global Warming to speed up as carbon levels show sharp rise, http://news.independent.co.uk/environment/article338689.ece, also available at: http://www.climateark.org/articles/reader.asp?linkid=50843

mately be reduced to through climate change is "a broken rabble led by brutal war-lords."[106]

The simplistic idea that humans could do as they wished with nature without substantially affecting their environment — at least beyond the local area — was never really correct. However, economic activities are now rapidly destabilizing the natural feedback loops that have maintained relative temperature and climatic stability throughout the entire ten thousand year span of human civilization. For example, tornado season in the US opened dramatically in March of 2006:

> ...reports show as many as 113 tornadoes may have hit five states on Sunday, marking a furious start to the Midwestern tornado season....meteorologists say the series of damaging storms will beat recent records for the number of tornadoes reported on a single day in March... Whatever the number, the figure still is likely to beat the [all time historical] record for the number of tornadoes to occur on a single day in March. The previous record was 59 on March 13, 1990.[107]

We're approaching the crisis-attractor ever faster, the "thrusters" are firing ever more rapidly; yet from our vantage on the "flight deck" all still appears reasonable normal.

Having sown the winds, we can now anticipate reaping the whirlwinds — not in some far off time, but sooner rather than later.

REAPING THE WHIRLWIND

During the 2005 hurricane season a commentator on television stated that "A hurricane is just nature's way of moving hot air to cooler regions." Looking at tropical storms in this way we must conclude that 2005's record-breaking hurricane season, which coincided with 2005 being the hottest year ever recorded, is a harbinger of what is to come.

The year 2005 saw the most named storms, the greatest number of hurricanes, a record number of major (category 3 and above) hurricanes hitting the US, and a record number of storms occurring before August 1st. In fact, records were broken or at least tied in virtually every tropical storm category. More evidence that 2005 was not a fluke is provided by multiyear cumulative statistics:[108]

106. *The Independent*, Jan. 16, 2006, Environment in crisis: "We are past the point of no return", http://news.independent.co.uk/environment/article338878.ece

107. Kansas.com, *The Wichita Eagle*, Mar. 14, 2006, Tornadoes Set Records for March, http://www.kansas.com/mld/kansas/14093208.htm

Tropical Storms	2-Year Consecutive Total: 42
	(previous record: 32, in 1995-96)
	3-Year Consecutive: 58
	(previous record: 43, most recently in 2002-04)
Total Hurricanes	2-Year Consecutive: 24
	(previous record: 21, in 1886-87)
	3-Year Consecutive: 30
	(previous record: 27, in 1886-88)
Total Major Hurricanes	2-Year Consecutive: 13
	(ties record in 1950-51)
	3-Year Consecutive: 16
	(ties records in 1949-51 and 1950-52)
Major Hurricane Landfalls	2-Year Consecutive: 7
	(previous record: 5, in 1954-55)
Florida Major Hurricane Landfalls	2-Year Consecutive: 5
	(previous record: 3, in 1949-50)

Hurricane strengths are classified by their core atmospheric pressure, in milibars (mb), with a lower core pressure in the hurricane's eye indicating a more energetic vortex. The year 2005 saw the strongest hurricane ever recorded, Hurricane Wilma at 882 mb. It also featured three of the six strongest hurricanes on record: Wilma at 882 mb (1st hurricane of the season), Hurricane Rita at 897 mb (4th), and Hurricane Katrina at 902 mb (6th).

A paper published in the 4 August 2005 edition of *Nature* entitled "Increasing Destructiveness of Tropical Cyclones over the Past 30 Years" along with another study published in the September 16, 2005 edition of *Science* entitled "Changes in Tropical Cyclone Number, Duration, and Intensity in a Warming Environment" demonstrated quantitatively that there had been a near doubling of total power incorporated in hurricanes, during a hurricane season, during the past 30 years.[109] These changes are attributed to global warming. However, the *Nature* study concludes that the increase in hurricane intensity far exceeds that predicted by climate models, and that, as warming increases, so too will the rapid rise in destructiveness of hurricanes.

---

108. National Oceanic and Atmospheric Administration, Nov. 29 2005, Noteworthy Records of the 2005 Atlantic Hurricane Season, http://www.noaanews.noaa.gov/stories2005/s2540b.htm; and Dr. Jeff Master's Wunder Blog, Jan. 6 2006, The Hurricane Season of 2005 Finally Ends!, http://www.wunderground.com/blog/JeffMasters/comment.html?entrynum=278&tstamp=200601

109. Changes in Tropical Cyclone, Number, Duration and Intensity in a Warming Environment, *Science* 309(5742): 1844-1846, 16 Sept., 2005, http://www.Sciencemag.org; and Increasing Destructiveness of Tropical Cyclones Over the Past 30 Years, *Nature* 436: 686-688, 4 Aug. 2005, http://www.nature.com

Another study published in *Geophysical Research Letters* on 12 August 2005, entitled *"The First South Atlantic Hurricane: Unprecedented Blocking, Low Shear and Climate Change"* found that the conditions which allowed the first ever South Atlantic hurricane to form were not attributable to random, improbable, circumstances. Rather, the study's authors demonstrated that climate change induced by global warming was affecting wind shear and other atmospheric condition in the South Atlantic in such a manner as to, for the first time, allow for hurricanes to form in that region.[110]

Two of 2005's Category 5 hurricanes, Rita and Katrina, tore through major oil production regions in the Gulf of Mexico as illustrated just below:

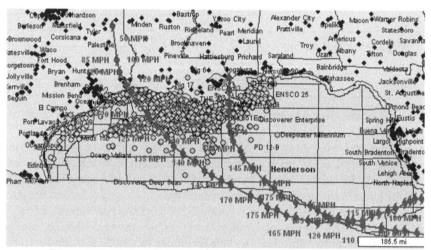

Figure 8. Hurricane Katrina and Rita Paths and Gulf Oil Platforms.
Source: RigLogix[111]

As a consequence of these storms, all production of oil and natural gas from the Gulf of Mexico was halted for weeks. As of late December 2005, official US government statistics indicated that 28 percent of oil production and 22.9 percent of natural gas production remained "shut-in," meaning that those percentages of production had not been restored.[112] By February of 2006, the government's Minerals Management Service (MMS) was reporting that 24 percent of oil production and about 16 percent of gas production continued to be "shut in."

110. The First South Atlantic Hurricane: Unprecedented Blocking, Low Shear, and Climate Change, *Geophysical Research Letters* 32 (L15712), 12 Aug. 2005, http://www.agu.org/journals/gl/

111. RigLogix – Hurricane Rita Interactive Map, http://gom.rigzone.com/rita.asp

112. US Energy Information Administration, Dec. 27 2005, Hurricane Impacts on the US Oil and Natural Gas Markets, http://tonto.eia.doe.gov/oog/special/eial_katrina.html

More ominously, MMS conceded that: "In the last few days there has been minimal improvement in the production numbers and this appears to be a trend that will continue with incremental movement over the next several months."[113] In other words, most of the remaining offline production facilities would still be offline at the onset of the 2006 hurricane season on June 1, 2006. Labor shortages attributable to these hurricanes are further hindering recovery.[114]

Finally, in March of 2006, Energy Secretary Sam Bodman told members of the House Energy and Commerce Committee that "Much of the oil and natural gas production still shut-in after last year's hurricanes in the Gulf of Mexico will stay offline because it would not be economical for companies to rebuild the production platforms."[115] According to MMS statistics, about one quarter of pre-hurricane oil production in the Gulf of Mexico remains "shut in" along with about 14% of natural gas production. Much of this offline production will never be rebuilt![116]

The hurricanes also severely damaged a considerable number of refineries in Louisiana and Texas:

> Hurricanes Katrina and Rita damaged a number of natural gas processing facilities on the Gulf Coast. The loss has and will continue to delay recovery of natural gas production in the area. Even if platforms and pipelines are either unaffected or readily restored to service, the gas often can't flow to market without treatment. In 2003 (the latest year with complete data), almost three-fourths of total US marketed gas production was processed prior to delivery to market. A number of processing plants in Louisiana and Texas, with capacities equal to or greater than 100 MMcf/d, are not active.[117]

Hurricane season 2006 opened early with tropical storm Alberto forming around June 10. By June 12 as Alberto approached the Florida panhandle, it briefly flirted with hurricane status before weakening decisively back to a tropical storm. For the Atlantic and Gulf of Mexico the 2006 hurricane season appears to be less intense that those of the past several years. Conversely, for the Pacific 2006 has been one of the most intense hurricane seasons ever recorded. Weather effects in the tropical Pacific — in particular the formation of "El Niño" conditions — appear to account for these outcomes. The Atlantic and Gulf of Mexico have been consistently warmer than their historical average throughout

---

113. The News Room [Minerals and Management Service Press Release # 3465] Feb. 8, 2006, Hurricane Katrina/Hurricane Rita Evacuation and Production Shut-in Statistics Report as of Feb. 8, 2006, http://www.mms.gov/ooc/press/2006/press0208a.htm

114. *Signonsandiego.com*, Dec. 20, 2005, Recovery of oil and gas output in Gulf, slowed by labor shortage, extent of damage, http://www.signonsandiego.com/news/business/20051220-1221-gulfoilrecovery.html

115. *Houston Chronicle*, March 9, 2006, Many Damaged Gulf Oil Rigs Won't Be Repaired, http://www.chron.com/disp/story.mpl/business/3712426.html

116. Minerals Management Service, The News Room, Press Release # 3479, Feb. 22, 2006, updated on March 8, 2006, http://www.mms.gov/ooc/press/2006/press0308b.htm

117. US Energy Information Administration, Dec. 27, 2005, Hurricane Impacts on the US Oil and Natural Gas Markets, http://tonto.eia.doe.gov/oog/special/eial_katrina.html

2006 — providing plenty of "fuel" for hurricane formation. However, other required atmospheric conditions needed for hurricane formation seem to have been temporarily disrupted in this region by the formation of El Niño conditions in the Pacific. Remember that in a system no one change can ever come without engendering other changes. Persistently warmer than average waters translate into persistently stronger hurricanes, on average.[118]

Logically, increasing warming must translate into more and more powerful storms in the years ahead. As has been noted, warmer ocean waters mean additional "fuel" for hurricanes in future years. The multi-year hurricane statistics provided above make this relationship between warming and storms clear; the likelihood of fresh major hurricanes striking the Gulf in 2006 and annually thereafter is much increased. We are at, or very near, the point in time when further investment in deepwater oil and natural gas facilities in the Gulf of Mexico becomes first economically impractical and then climatologically impossible.

This observation is confirmed by the findings of a study published in June 2006:

> Global warming accounted for around half of the extra hurricane-fueling warmth in the waters of the tropical North Atlantic in 2005, while natural cycles were only a minor factor, according to a new analysis by Kevin Trenberth and Dennis Shea of the National Center for Atmospheric Research (NCAR). The study will appear in the June 27 issue of *Geophysical Research Letters*, published by the American Geophysical Union....Global warming does not guarantee that each year will set records for hurricanes, according to Trenberth. He notes that last year's activity was related to very favorable upper-level winds as well as the extremely warm SSTs. Each year will bring ups and downs in tropical Atlantic SSTs due to natural variations, such as the presence or absence of El Nino, says Trenberth. However, he adds, the long-term ocean warming should raise the baseline of hurricane activity.[119]

Further confirmation came in July 2006 when a CNN report citing the *Wall Street Journal* stated that:

> Oil rigs are leaving the Gulf of Mexico in record numbers, threatening to put upward pressure on US oil and natural gas prices, according to a report published Wednesday. Drilling companies are increasingly signing long-term deals with oil firms to send their rigs to more promising drilling regions overseas, said the *Wall Street Journal*.[120]

---

118. Weather Underground, Dr. Jeff Master's WunderBlog, El Niño is Coming, http://www.weatherunderground.com/blog/JeffMasters/comment.html?entrynum=506&tstamp=200609

119. Physorg.con, Science, Technology, Physics, Space News, Space and Earth Science, June 22, 2006, Global Warming Surpassed Natural Cycles in Fueling 2005 Hurricane Season, http://www.physorg.com/news70203350.html

120. *CNNMoney.com*, July 5, 2006, Oil rigs leaving Gulf of Mexico, Drilling companies are sending vital shallow water drilling-rigs abroad, driving up domestic energy prices, according to a report, http://money.cnn.com/2006/07/05/news/economy/oil_rigs/index.htm

The report noted that the lost natural gas production from the Gulf could not be made up as the US has little to no ability to increase its imports of natural gas. Thus, for supply to balance with demand, prices must rise.

Of course, accelerating climate change makes the US's need for oil and natural gas ever more desperate. This growing need then causes the political leadership — primarily drawn from, or funded by, oil companies to begin with — to press ever more aggressively to implement two strategies described in Chapter 3:

(1) Use of military power abroad to secure control of essential energy supplies and transit routes.

(2) Desperate crash programs to produce oil substitutes domestically using coal, oil shale and oil sands. This will occur in conjunction with rapidly increasing use of nuclear power.

Both of these strategies amount to fiddling while Rome burns. Political collapse and global warming are further accelerated by the use of these two strategies. And the two strategies accelerate one another in positive feedback loops: Vast quantities of coal are rapidly burned with no environmental safeguards whatsoever, to prop up collapsing economic and political systems; this burning releases ever greater amounts of greenhouse gases ever more rapidly into the atmosphere. Global warming accelerates. Climate collapse occurs, which increases desperation on the part of political and economic elites still more.

No one knows if, right now, global warming has indeed become unstoppable — yet. However, there can be no question that if it has not yet become so, it soon must. Unless, of course, the major shareholders of the major corporations which control most meaningful decision-making by the planet's major nation-states were to altruistically abandon much of those activities from which their wealth and power are derived. And that simply will not happen.

And so runaway global warming seems inevitable. It is possible that other unanticipated negative feedback loops could occur, such as increased temperature causing greater cloud cover, leading to at least some additional sunshine being reflected out to space, producing some net cooling. Regardless, atmospheric CO_2 is itself a powerful greenhouse gas; so once again, we see that increased warming — runaway warming — is quite likely in our very near future.

Scientific discussions of environmental change and global warming have long been haunted by the specter of nonlinearity. Climate models, like econometric models, are easiest to build and understand when they are simple linear extrapolations of well-quantified past behavior; when causes maintain a consistent proportionality to their effects. But all the major components of global climate — air, water, ice, and vegetation — are actually nonlinear: At certain thresholds they can switch from one state of organization to another, with catastrophic consequences for species too finely-tuned to the old norms. Until the early 1990s, however, it was generally believed that these major climate transitions took centuries, if not millennia, to accomplish. Now, thanks to the decoding of subtle signatures in ice cores and sea-bottom sediments, we know that global temperatures and ocean circulation can, under the right circumstances, change abruptly — in a decade or even less....Abrupt switching mechanisms in the climate system — such as relatively small changes in ocean salinity — are augmented by causal loops that act as amplifiers. Perhaps the most famous example is sea-ice albedo: The vast expanses of white, frozen Arctic Ocean ice reflect heat back into space, thus providing positive feedback for cooling trends; alternatively, shrinking sea-ice increases heat absorption, accelerating both its own further melting and planetary warming. Thresholds, switches, amplifiers, chaos — contemporary geophysics assumes that Earth history is inherently revolutionary.[121]

Potential effects of rapid climate change extend far beyond unseasonably warm weather:

> Climate change could lead to the extinction of many animals including migratory birds, says a report commissioned by the UK government.... the fear is that the changes currently under way are simply too rapid for species to evolve new strategies for survival. Their options are also being narrowed by the rapid conversion of ecosystems such as the draining of wetlands, felling of forests and development of coastlines — so if their existing habitats are hit by global warming, there is literally no place to go.[122]

Consider again that "...we know that global temperatures and ocean circulation can, under the right circumstances, change abruptly — in a decade or even less." One of our ingrained expectations about reality is that even if global climate change does occur, it will be gradual. No belief could be more false — and consequentially more dangerous!

In the United Kingdom the Oxford Research Group, a "think tank," has just released a report which found that the effects of climate change will present far greater threats to the security and order of the planet than will religiously or politically motivated terrorism. The resulting food shortages and social destabilization are likely to trigger deepening resentment of affluent nations by poorer ones, leading to greater international violence, terrorism, and the general breakdown of order across the planet. The report's summary states that:

> The issues analysed in this report are those that are likely to dominate the international security environment over the next 30 years. Unless urgent action is taken in the next five to ten years, it will be extremely difficult, if not impossible, to avoid a highly unstable global system by the middle years of the century. Governments, NGOs [Non-Governmental Organizations] and concerned citizens must work together and recognise that they now have an urgent responsibility to embrace a sustainable approach to global security.[123]

---

121. *Common Dreams News Center*, Oct. 7, 2005, The Other Hurricane: Has the Age of Chaos Begun?, http://www.commondreams.org/views05/1007-20.htm
122. *BBC News International*, Oct. 5, 2005, Animals 'hit by global warming', http://news.bbc.co.uk/2/hi/science/nature/4313726.stm

Increasingly rapid climate change will increasingly unleash political and economic convulsions across the planet. Fundamental climate change is, by itself, sufficient to collapse the global political economy. Think of weather disruptions on the level of New Orleans in summer 2005 occurring across the planet regularly.

A major reason for the empirically observed increases in "extreme weather" seems to be clear. Weather is simply the result of uneven heating of the atmosphere and oceans. Weather is how areas of higher temperature (energy) flow towards areas of lower temperature (energy) to bring about equalization of temperature (energy). The density of the ocean is about one hundred times that of the atmosphere, and it follows that far more heat equalization is brought about by oceanic currents than by atmospheric phenomena.

However, the global oceanic current is now slowing down due to increasing infusions of melt from freshwater glaciers. The magnitude of this slowdown is vast. An article published in the December 1, 2005 edition of *Nature* entitled "*Slowing of the Atlantic Meridional Overturning Circulation at 25 Degrees North*"[124] shows that the reduction in flow for the Gulf Stream, measured between 1998 and 2004, at one point in the North Atlantic is equal to 60 times the annual flow of the Amazon River. This means that a truly vast quantity of heat energy is no longer able to be circulated from warm areas to cold ones.

Yet the temperature differential between these regions still exists. Therefore, the atmosphere must serve as the heat transfer medium. We humans experience this greatly increased use of the atmosphere to move energy from high to low potential — from hot areas to cooler ones — as an increased likelihood of violent storms. Hurricanes, tornadoes, torrential rains, thunderstorms, higher winds and etc., are all manifestations of excess energy diffusing via the atmosphere rather than by means of oceanic currents. Again, no cruel gods are tormenting us. Rather, there is only energy behaving in accord with fundamental laws of physics.

123. Oxford Research Group, June 2006, Global Responses To Global Threats, Sustainable Security For The 21st Century, http://www.oxfordresearchgroup.org.uk/publications/briefings/globalthreats.pdf, pp 32.
124. Bryden, Harry L. et al., "Slowing of the Atlantic Meridional Overturning Circulation at 25 Degrees North", *Nature* 438: 655-657, 1 Dec., 2005, http://www.nature.com

CHAPTER 5. GLOBAL WARMING AND CLIMATE CHANGE II: OCEANS, WATERS, AND ICES

There are no passengers on spaceship earth. We are all crew.
— Marshall McLuhan[125]

ICECAPS AND OCEANS

Seven-tenths of the earth's surface is covered by water. Its capacity for heat storage, for carbon storage, and its ability to affect the planet's overall climate, is much greater than is the case for any other part of the planetary life support system.[126]

Paradoxically, because of its effects upon ocean currents, global warming can cause regional cooling.

Great Britain and northwestern Europe are about nine degrees Fahrenheit warmer than they otherwise would be, due to the heat conveying effects of the Atlantic Ocean's Gulf Stream.[127] The process works like this: warm ocean waters flow northwards across the equator towards Europe. Along the way, they are heated by the tropical sun. As they flow past northern Europe, they release this accumulated thermal energy as heat. The current circles Iceland where it turns west and then south flowing past Greenland. The high concentration of salt in the water under the Greenland ice sheet causes the cooler heat-depleted

---

125. Quoteland.com: Marshall McLuhan http://www.quoteland.com/author.asp?AUTHOR_ID=295
126. NASA, Earth Observatory, Reference: Ocean and Climate, "The Earth's Ocean and Atmosphere are Locked in an Embrace — as One Changes So Does the Other", http://earthobservatory.nasa.gov/Library/OceanClimate/
127. Masters, Jeff, The Weather Underground, inc, The Science of Abrupt Climate Change, http://www.wunderground.com/education/abruptclimate.asp

water to sink downwards into the Greenland Sea, making it the Gulf Stream's "pump." As this water sinks towards to the bottom of the ocean, it is replaced by warm water flowing in from the south. And the cycle repeats itself.

And so day after day, year after year, the heavily-populated, economically-developed quadrant of northwest Europe is consistently heated by this steady Gulf flow to temperatures well above what they would otherwise be. A glance at a map of world oceanic currents illustrates this clearly:

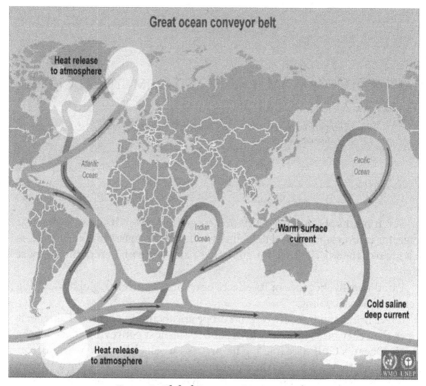

Figure 1. Global Oceanic Conveyor Belt.

Source: UN Intergovernmental Panel on Climate Change reproduced in Weather Underground[128]

The latitudes of Britain and northern Europe correspond to those of northern Canada and Greenland! Latitudes which on the American side of the Atlantic are frozen arctic expanses are temperate and fertile on the European side. That is the effect of the Gulf Stream!

---

128. Ibid., Masters, # 127

This raises the question of just how stable the heating produced by this Gulf Stream current might be. Dr Jeff Masters writing for the on-line weather information site Weather Underground comments that:

> In the tropical Atlantic, the sun's heat evaporates large amounts of water, creating relatively warm, salty ocean water. This warm, salty water flows westward toward North America, then up the East Coast of the U.S, then northeastward toward Europe, forming the mighty Gulf Stream current. As this warm, salty water reaches the ocean regions on either side of Greenland, cold winds blowing off of Canada and Greenland cool the water substantially. In [the Figure just above] these regions are marked with white circles labeled, "Heat release to the atmosphere." These cool, salty waters are now very dense compared to the surrounding waters, and sink to the bottom of the ocean. Thus, the oceanic areas by Greenland where this sinking occurs are called "deep-water formation areas". This North Atlantic deep water flows southward toward Antarctica, eventually making it all the way to the Pacific Ocean, where it rises back to the surface to complete the Great Ocean Conveyor Belt. It takes about 1000 years for the water to make a complete circuit around the globe. Since the Great Ocean Conveyor belt is driven in part by differences in ocean water density, if one can pump enough fresh water into the ocean in the key areas on either side of Greenland where the Gulf Stream waters cool and sink, this will lower the ocean's salinity (and therefore its density) enough so that the waters there no longer sink. The Atlantic conveyor belt and Gulf Stream current will then shut down in just a few years, dramatically altering the climate.[129]

This means that if the glaciers which cover almost all of Greenland to a depth of up to two miles were to begin to melt rapidly, the resulting influx of fresh water would reduce the salinity of the ocean water, preventing it from sinking as it grows cold, and likely shutting down the Gulf Stream current virtually overnight. Furthermore, the entire global oceanic conveyor belt would be massively disrupted across the planet with incalculable effects everywhere.

Is this likely? In 1993, separate teams of American and European researchers working about 20 miles apart succeeded in drilling completely through two miles of ice sheets in Greenland. The ice cores they thereby acquired provided a record of northern hemispheric climate during the past 110,000 years. These findings proved to be absolutely astounding:

> What the scientists found was surprising and unnerving. Scientists had known from previous ice core and ocean sediment core data that Earth's climate had fluctuated significantly in the past. But what astonished scientists was the rapidity with which these changes occurred. As seen in Figure 1, the ice core record showed *frequent* sudden warmings and coolings of 15°F (8°C) or more. Many of these changes happened in less than 10 years. And in at least one case 11,600 years ago, when Earth emerged from the final phase of the most recent ice age (an event called the Younger Dryas), the Greenland ice core data showed that a 15°F (8°C) warming occurred in less than a decade, accompanied by a doubling of snow accumulation in 3 years. Most of this doubling occurred in *a single year*. Ocean and lake sediment data from places such as California, Venezuela, and Antarctica have confirmed that these sud-

---

129. Ibid., Masters # 127

den climate changes affected not just Greenland, but the entire world. And during the past 110,000 years, there have been at least 20 such abrupt climate changes. Only one period of stable climate has existed during the past 110,000 years — the 11,000 years of modern climate (the "Holocene" era). "Normal" climate for Earth is the climate of sudden extreme jumps — like a light switch flicking on and off.[130]

These findings are graphically represented in Figure 2:

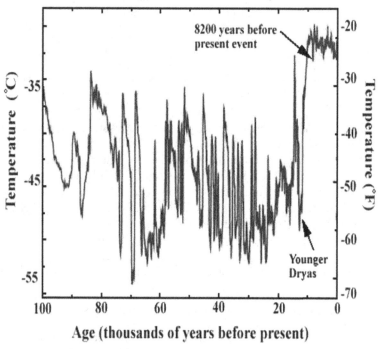

Temperatures in Greenland over the past 100,000 years

Age (thousands of years before present)

Figure 2. Temperatures in Greenland over the Past 100,000 Years.

Source: *Journal of Geophysical Research, 102, 383-396, 1997. Reproduced in Weather Underground*[131]

130. Ibid., Masters # 127
131. Cuffey, K.M, and G.D. Clow, "Temperature, accumulation, and ice sheet elevation in central Greenland throughout the last deglacial transition", *Journal of Geophysical Research*, 102, 383-396, 1997. Reproduced in http://www.wunderground.com/education/abruptclimate.asp

In other words, the climatic stability of the modern era is an aberration and not only that, but rapid, nearly instantaneous climate shifts due to the shutdown of the Gulf Stream conveyor belt are historically common. This fact, established by ice core samples, was noted in a report presented to the American Geophysical Union meeting in 2004 by two researchers, Michael Schlesinger and Natasha Andronova:

> Paleoclimate records constructed from Greenland ice cores have revealed that the thermohaline circulation has, indeed, shut down in the past and caused regional climate change. As the vast ice sheet that covered much of North America during the last ice age finally receded, the meltwater flowed out the St. Lawrence and into the North Atlantic.

> "The additional fresh water made the ocean surface less dense and it stopped sinking, effectively shutting down the thermohaline circulation," Schlesinger said. "As a result, Greenland cooled by about 7 degrees Celsius within several decades. When the meltwater stopped, the circulation pattern restarted, and Greenland warmed."[132]

This presses the question as to the likelihood that today's process of global warming will accelerate the melting of the Greenland icecap, thereby triggering the shutting down of the Gulf current and causing an abrupt climate change — a localized ice age — in the European North Atlantic, and whatever follow-on effects that would have. Until recently, researchers had been divided and unsure about this. However, recent research appears to have answered this question. Several multi-year studies relying upon actual measurements of the Gulf Stream were published during 2005. Their findings were consistent. First this:

> The powerful ocean current that bathes Britain and northern Europe in warm waters from the tropics has weakened dramatically in recent years, a consequence of global warming that could trigger more severe winters and cooler summers across the region, scientists warn today.

> Researchers on a scientific expedition in the Atlantic Ocean measured the strength of the current between Africa and the east coast of America and found that the circulation has slowed by 30% since a previous expedition 12 years ago.[133]

And then this:

> Cambridge University ocean physics professor Peter Wadhams points to changes in the waters of the Greenland Sea. Historically, large columns of very cold, dense water in the Greenland Sea, known as "chimneys," sink from the surface of the ocean to about 9,000 feet below to the seabed. As that water sinks, it interacts with the warm Gulf Stream current flowing from the south. But Wadhams says the number of these "chimneys" has dropped from about a dozen to just two. That is causing

---

132. *Science Daily*, 20 Dec. 2004, Shutdown of Circulation Pattern Could be Disastrous Researchers Say, http://www.sciencedaily.com/releases/2004/12/041219153611.htm.

133. *Guardian Unlimited*, Dec. 1, 2005, Alarm over dramatic weakening of Gulf Stream, http://www.guardian.co.uk/science/story/0,3605,1654803,00.html?gusrc=rss

a weakening of the Gulf Stream, which could mean less heat reaching northern Europe. The activity in the Greenland Sea is part of a global pattern of ocean movement, known as thermohaline circulation, or more commonly the "global conveyor belt."[134]

Rapid melting of the Greenland icecap, with all that this implies for the climate of Europe, is indeed underway:

> Greenland's glaciers have begun to race towards the ocean, leading scientists to predict that the vast island's ice cap is approaching irreversible meltdown, *The Independent on Sunday* can reveal. Research to be published in a few days' time shows how glaciers that have been stable for centuries have started to shrink dramatically as temperatures in the Arctic have soared with global warming. On top of this, record amounts of the ice cap's surface turned to water this summer. The two developments — the most alarming manifestations of climate change to date — suggest that the ice cap is melting far more rapidly than scientists had thought, with immense consequences for civilisation and the planet.[135]

This finding is further reinforced by testimony given by Dr. Robert W. Corell in testimony before Congress:

> The likelihood of continued melting of glaciers including the Greenland Ice Sheet have significant implications for the entire planet as the total land-based ice in the Arctic has been estimated to be about 3,100,000 cubic kilometers which, if melted, would correspond to a sea-level equivalent of about 8 meters. The Greenland Ice Sheet dominates land ice in the Arctic. Over the past two decades, the melt area on the Greenland ice sheet has increased on average by about 0.7%/year (or about 16% from 1979 to 2002), with considerable variation from year to year [See Figure on the next page.] The total area of surface melt on the Greenland Ice Sheet broke all past records in 2002.[136]

This process of disintegration may occur much more rapidly than current models suggest. Dr. Jim Hansen is the director of NASA's Goddard Institute for Space Studies. He recently made news when it was revealed that the Bush Administration had attempted to muzzle him over the issue of how rapidly climate change was unfolding. Dr. Hansen states:

---

134. *CNN.com*, May, 10, 2005, Changes in Gulf Stream Could Chill Europe, http://www.cnn.com/2005/TECH/science/05/10/gulfstream/

135. *The Heat is Online*, Glaciers, http://www.heatisonline.org/Glaciers.cfm

136. Statement by Dr. Robert W. Corell, Chair Arctic Climate Impact Assessment Organized by the Arctic Council and the International Arctic Sciences Committee and Senior Fellow, American Meteorological Society before The Committee on Commerce, Science, and Transportation United States Senate March 3, 2004, http://www.groundtruthinvestigations.com/documents/climatechange.html

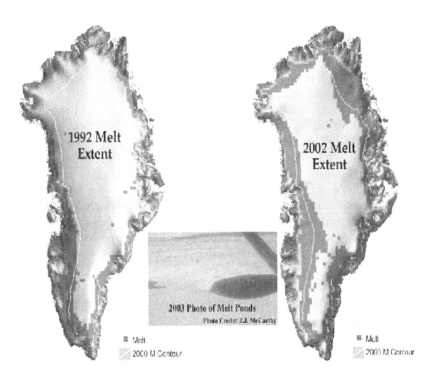

Figure 3. Increasing Melting of Greenland Icecap.

Source: Dr. Robert W. Corell, testimony before the Committee on Commerce, Science, and Transportation, United States Senate March 3, 2004.[137]

Once a sheet starts to disintegrate, it can reach a tipping point beyond which break-up is explosively rapid. The issue is how close we are getting to that tipping point. The summer of 2005 broke all records for melting in Greenland. So we may be on the edge. Our understanding of what is going on is very new. Today's forecasts of sea-level rise use climate models of the ice sheets that say they can only disintegrate over a thousand years or more. But we can now see that the models are almost worthless....But what is happening is much more dynamic. Once the ice starts to melt at the surface, it forms lakes that empty down crevasses to the bottom of the ice. You get rivers of water underneath the ice. And the ice slides towards the ocean....How fast can this go? Right now, I think our best measure is what happened in the past. We know that, for instance, 14,000 years ago sea levels rose by 20m in 400 years — that is five metres in a century....How far can it go? The last time the world was three degrees warmer than today — which is what we expect later this century — sea levels were 25m higher....None of the current climate and ice models predict this. But I prefer the evidence from the earth's history and my own eyes....It's hard to say what the world will be like if this happens. It would be another planet....How long have we got? We have to stabilise emissions of carbon dioxide within a decade, or temperatures will warm by more than one degree. That

---

137. Ibid., Corell, # 136

will be warmer than it has been for half a million years, and many things could become unstoppable.[138]

Dr. Hansen estimates we have no longer than a decade to stabilize carbon dioxide emissions before warming and melting and all of their associated effects become "unstoppable."

Bush will remain in office until 2009. Dr. Hansen issued his ultimatum in 2005. Half of this final decade will have elapsed before Bush leaves office. The eight-year term (2001 — 2009) of the G.W. Bush Administration may have represented the final opportunity humanity had to come to terms with global warming — and peak oil, too, before collapse became unavoidable. That opportunity, obviously, is not being taken.

For Ireland, Britain and all of northern Europe, the sudden transition to Siberian weather, brought about by the Greenland icecap's melting, would mean the end of their world. Their lands will become virtually uninhabitable. This disaster alone without taking into consideration the rippling, worldwide climatic effects of the global conveyor belt's disruption, much less all of the other impending crises, is sufficient to bring our global economy crashing down.

Working on both sea and land, nature itself appears to be intervening to hasten this process:

> British researchers, examining almost six thousand soil borings across the UK, found another feedback effect. Warmer temperatures (growing seasons now last eleven days longer at that latitude) meant that microbial activity had increased dramatically in the soil. This, in turn, meant that much of the carbon long stored in the soil was now being released into the atmosphere. The quantities were large enough to negate all the work that Britain had done to switch away from coal to reduce carbon in the atmosphere. "All the consequences of global warming will occur more rapidly," said Guy Kirk, chief scientist on the study. "That's the scary thing. The amount of time we have got to do something about it is smaller than we thought."[139]

Of course, this finding is true across the planet, not just in Britain. Since existing computer models do not incorporate this newly discovered positive feedback loop, they are inaccurate. The forces now accelerating the process of global warming are greater than we have yet accounted for.

In research published in *Nature* in 2006, a scientific team led by Dr. Jef Huisman of the University of Amsterdam demonstrates that global warming produces much warmer surface waters in the world's oceans. This in turn creates a separation between these waters and deeper waters, which remain

---

138. *The Independent*, 19 Feb. 2006, Climate Change: On the Edge, Greenland Icecap Breaking up at Twice the Rate it was Five Years Ago Says Scientist Bush Tried to Gag, http://news.independent.co.uk/environment/article345926.ece

139. *The New York Review of Books*, Volume 53, Number 1, Jan. 12, 2006, The Coming Meltdown, http://www.nybooks.com/articles/18616 Ref. paragraph 4

cooler. They occupy two different bands of water, which do not mix significantly as they did in pre-global warming times.

> [G]lobal warming, which is causing the temperature of the sea surface to rise, will also interfere with the vital upward movement of nutrients from the deep sea. These nutrients, containing nitrogen, phosphorus and iron, are vital food for phytoplankton. If the supply is interrupted the plants die off, which prevents them from absorbing carbon dioxide from the atmosphere. Global warming of the surface layers of the oceans reduces the upward transport of nutrients into the surface layers. This generates chaos among the plankton," the professor [Huisman] said. The sea is one of nature's "carbon sinks," which removes carbon dioxide from the atmosphere and deposits the carbon in a long-term store — dissolved in the ocean or deposited as organic waste on the seabed.[140]

The research conducted by Dr. Huisman and his scientific team began with the creation of a detailed computer simulation. They then tested their model's predictions against actual data taken from the Pacific Ocean. Their findings were in close accord with their model's predictions. These findings are hugely significant for two reasons:

Phytoplankton forms the base of the entire oceanic food chain. As their numbers crash, so too must the entire oceanic ecosystem.

Phytoplankton store huge amounts of excess CO_2 in their bodies, thus keeping it out of the atmosphere. A sudden decline in their numbers means an equally sudden increase in the rate at which CO_2 accumulates in the atmosphere. This represents a positive feedback loop, which rapidly increases the rate of increase of CO_2 in the atmosphere; therefore, the rate at which the world warms must also increase ever faster. The net result is that the rate of global warming increases still more rapidly, even as marine fisheries collapse producing worldwide famine — something that will be discussed in the next chapter.

This warming triggers still other positive feedback mechanisms which, in turn, further accelerate the warming:

> Arctic sea ice is melting fast. There was 20 percent less of it than normal this summer, and as Dr. Mark Serreze, one of the researchers from Colorado's National Snow and Ice Data Center, told reporters, "the feeling is we are reaching a tipping point or threshold beyond which sea ice will not recover." That is particularly bad news because it creates a potent feedback effect: instead of blinding white ice that bounces sunlight back into space, there is now open blue water that soaks up the sun's heat, amplifying the melting process.[141]

The Earth's other polar region is experiencing rapid net warming as well. A slew of recent studies and experiments have produced complementary and self-

---

140. The Independent, 19 Jan. 2006, Warmer seas will wipe out plankton, source of ocean life, http://news.independent.co.uk/environment/article339596.ece Article can also be found at: http://www.truthout.org/issues_06/011906EC.shtml

141. Ibid., *New York Review of Books*, # 138, Ref. paragraph 2

consistent findings. What they are portraying is a panorama of a continent in which rapid net warming is producing accelerating melting of glacial ice.

In recent decades the Antarctic Peninsula, located in northernmost portion of the continent, has experienced mean temperature increases of 2.5 degrees Celsius or 4.5 degrees Fahrenheit. This has already caused the 770 square mile (2,000 square kilometer) Larsen A Ice Shelf to abruptly disintegrate in early 2002. This follows the breakup of the 1,150 square mile (3,000 square kilometer) Larsen B and Wilkins Ice Shelves between March of 1998 and March of 1999.

Signs of ecological stress are increasing. Adelie penguin populations have already decreased by about one third due to reductions in winter sea ice which they inhabit. This reduction in sea ice and its associated ecology, upon which these penguins rely on for food, has increasingly devastated the entire regional marine ecology:

> Climate change and disappearing sea ice in the Southern Ocean are causing food shortages that could threaten Antarctic whales, seals and penguins, scientists say. The vanishing ice in the winter has resulted in an 80% drop in the number of Ant-arctic krill, a shrimp-like crustacean that is a major source of food for animals in the region. "This is the first time that we have understood the full scale of this decline," said Dr Angus Atkinson, a marine biologist at the British Antarctic Survey, who reported the research in today's [Nov. 4, 2004] issue of the journal *Nature*. Krill feed on algae under the ice sheet in the ocean but warmer temperatures over the past 50 years have meant there is less ice and fewer krill.[142]

These new findings are particularly significant because previous studies had projected that increased snowfall would result from warming temperatures adding more moisture to the air, resulting in more snow falling. This negative feedback effect was expected to increase the mass of Antarctic ice and snow faster than warming temperatures could melt it along its edges. However, for Antarctica, melting clearly overwhelmed the conflicting effect of increasing snowfall. Antarctica is losing the equivalent of 36 cubic miles of ice each year due to net melting.

In the case of Greenland, the study showed a slight net gain for the ice sheet between 1992 and 2002. However, a more recent study analyzing data through 2005 indicates that melting has now overtaken increased snowfall there too.[143] By 2005 the net melting had reached 52 cubic miles per year and was accelerating, even as the Greenland ice sheet showed increasing signs of fis-suring.[144]

---

142. *News in Science*, Nov. 4, 2004, Antarctic Warming Killing Fish Food, http://abc.net.au/science/news/stories/s1234914.htm

143. NASA.gov, Impact of Climate warming on Polar ice sheets Confirmed, March, 8, 2006, http://www.nasa.gov/vision/earth/environment/ice_sheets.html; and *The New York Times*, March, 3, 2006, Loss of Antarctic Ice Increases, http://www.nytimes.com/2006/03/03/science/03melt.html?_r=1&oref=login

Overall, there was a net melting of freshwater from the Antarctic and Greenland ice caps of about 20 billion tons per year. And the loss is, very clearly, accelerating.

Some sense of the enormous stress climate change is placing upon polar ecosystems in general, and specifically upon the animals trying to survive within these rapidly changing ecosystems, can be gleaned from a recent international scientific study of polar bear behavior in the North American arctic:

> Polar bears in the southern Beaufort Sea may be turning to cannibalism because longer seasons without ice keep them from getting to their natural food, a new study by American and Canadian scientists has found. The study reviewed three examples of polar bears preying on each other from January to April 2004 north of Alaska and western Canada, including the first-ever reported killing of a female in a den shortly after it gave birth. Polar bears feed primarily on ringed seals and use sea ice for feeding, mating and giving birth. Polar bears kill each other for population regulation, dominance, and reproductive advantage, the study said. Killing for food seems to be less common, said the study's principal author, Steven Amstrup of the US Geological Survey
>
> Alaska Science Center. "During 24 years of research on polar bears in the southern Beaufort Sea region of northern Alaska and 34 years in northwestern Canada, we have not seen other incidents of polar bears stalking, killing, and eating other polar bears," the scientists said. Environmentalists contend shrinking polar ice due to global warming may lead to the disappearance of polar bears before the end of the century.[145]

To understand what this warming means in human terms, consider how the native Inuit in Canada's far north are affected. These ice hunters have experienced one warm winter after another in recent years. The cumulative effects of this warming have been to make it ever more difficult for them to capture the diminishing populations of fish and game upon which they subsist. Further, warming weather has made much of the winter ice too thin to be traversable. Canada's Federal Weather Service has reported that Canada experienced its warmest winter on record in 2005-2006. For Canada as a whole, mean winter temperatures were 7 degrees Fahrenheit above historical average. The farther north the location, the greater the warming was. One Inuit elder named Metuq spoke particularly poignantly.

> Metuq, the hunter, fears the worst. "The world is slowly disintegrating," he said, inside his heated house in Pangnirtung, a community of 1,200 perched on a dramatic

---

144. *Los Angeles Times*, LATimes.com, June 25, 2006, Greenland's Ice Sheet is Slip Sliding Away, http://www.latimes.com/news/science/la-sci-greenland25jun25,0,1308610.story?page=1&coll=la-home-headlines

145. Associated Press reported in Yahoo News, June 12, 2006, Polar Bears May be Turning to Cannibalism, http://news.yahoo.com/s/ap/20060612/ap_on_sc/polar_bear_cannibalism;_ylt=AuFLhl4_8PNmz_lblGfsZ3WsONUE;_ylu=X3o DMTA3b2NibDltBHNlYwM3MTY-

union of mountain and fjord on Baffin Island. Seal skins stretched on canvas dried outside his home. The town remained treacherous. Rain in February had frozen solid, and there had been almost no snow to cover it. "They call it climate change," he said. "But we just call it breaking up." The troubles for the Inuit are ominous for everyone, says Sheila Watt-Cloutier, head of the International Circumpolar Conference, an organization for the 155,000 Inuit worldwide. "People have become disconnected from their environment. But the Inuit have remained through this whole dilemma, remained extremely connected to its environment and wildlife," she said. "They are the early warning. They see what's happening to the planet, and give the message to the rest of the world."[146]

It is almost eerie that the very words which Metuq chose to describe the rapid climate change he is experiencing, "breaking up" and "disintegrating" are the exact words which best describe the fate of the Columbia in my Prologue metaphor for contemporary human civilization. Similar stories to Metuq's abound in the high arctic:

> In Pangnirtung, residents were startled by thunder, rain showers and a temperature of 48 degrees in February, a time when their world normally is locked and silent at minus-20 degrees. "We were just standing around in our shorts, stunned and amazed, trying to make sense of it," said one resident, Donald Mearns. "These are things that all of our old oral history has never mentioned," said Enosik Nashalik, 87, the eldest of male elders in this Inuit village. "We cannot pass on our traditional knowledge, because it is no longer reliable. Before, I could look at cloud patterns or the wind, or even what stars are twinkling, and predict the weather. Now, everything is changed."[147]

Imagine temperatures 70 degrees Fahrenheit *above* average occurring ever more frequently.

In spring 2006 researchers from the British Antarctic Society published research indicating that air temperatures across then entire continent of Antarctica were warming dramatically at a rate three times greater than that of the rest of the planet:

> Although the Antarctic Peninsula has warmed by more than 2.5C during the past 50 years, most surface measurements suggest that there have been no pronounced temperature changes elsewhere on the continent, while some have indicated a small cooling effect. The new research, led by John Turner, of the BAS, shows that the air above the surface of Antarctica is definitely warming, in ways that are not predicted by climate models and that cannot yet be explained. The results are published today in the journal *Science*. "The rapid surface warming of the Antarctic Peninsula and the enhanced global warming signal over the whole continent shows the complexity of climate change," Dr Turner said. "Greenhouses gases could be having a bigger impact in Antarctica than across the rest of the world and we don't understand why. "The warming above the Antarctic could have implications for snowfall across

146. Inuit See Signs in Arctic Thaw, String of Warm Winters Alarms "Sentries for the Rest of the World", *Washington post.com*, March, 23, 2006, Page A-1, http://www.washingtonpost.com/wp-dyn/content/article/2006/03/21/AR2006032101722.html
147. Ibid., Inuit, # 146

the Antarctic and sea level rise. Current climate model simulations don't reproduce the observed warming, pointing to weaknesses in their ability to represent the Antarctic climate system. Our next step is to try to improve the models."[148]

This warming is occurring much more rapidly than current climatological models predict.

In mid-2006 geoscientist Allen Glazner announced the findings from his research into possible relationships between glacial melting and seismic and volcanic activity. He found that periods of rapid glacial melting corresponded with periods of unusual earthquake intensity and volcanism. Correlation does not prove causation; however, as explained by Sharon Begley, the science correspondent for the Wall Street Journal, Professor Glazner observes that "one cubic meter of ice weighs just over a ton, and glaciers can be hundreds of meters thick." He then suggests that "when they melt and the water runs off, it is literally a weight off earth's crust" where the ice used to be and an increase in pressure where the water goes....So that eruptions might increase where "glacial retreat lifts pressure that had kept the magma conduit closed."[149]

Prof. Glazner's findings have been corroborated by the research of a team of scientists working for NASA and the US Geological Survey:

> In the July Global and Planetary Change, Jeanne Sauber, a geophysicist at NASA's Goddard Space Flight Center in Greenbelt, Md., and Bruce Molnia, a geologist with the US Geological Survey in Reston, Va., report that in the past 80 years, the wastage of coastal glaciers in the Icy Bay and Malaspina regions of Alaska is so great that it has reduced the stability of faults in the region and may have hastened the magnitude-7.2 St. Elias quake in 1979. "I've learned that if you are going to be doing earthquake hazard assessments in places like southern Alaska," Sauber says, "you need to include the history of what glaciers are doing." Along the southern coast of Alaska, the Pacific plate subducts beneath the North American plate at a rate of roughly 50 millimeters per year. The resulting strain in the crust accumulates until an earthquake releases the pent-up energy. When large glaciers or ice sheets move across the landscape, their immense weight actually depresses Earth's crust. When glaciers retreat, the ground rebounds back into place. In tectonically active areas experiencing glacial retreat and postglacial rebound, "earthquakes may occur sooner" than if they were still covered by glaciers, Sauber says. Erik Ivins, a geophysicist at NASA's Jet Propulsion Laboratory in Pasadena, Calif., says that earthquakes that occur in places previously covered by glaciers may be larger than normal "because the ice mass overburden was keeping smaller earthquakes from releasing tectonic stress."...The relationship between earthquakes and glaciers in tectonically quiescent areas supports the research, says Thomas James, a geophysicist with Natural Resources Canada in British Columbia. In Scandinavia and Canada, large earth-

148. Times Online, Antarctic air is warming faster than rest of world, March, 31, 2006, http://www.timesonline.co.uk/article/0,,3-2111772,00.html

149. The Wall Street Journal Online, June 9, 2006, How Melting Glaciers Alter Earth's Surface, Super Quakes, Volcanoes, http://online.wsj.com/public/article/SB114981650181275742sOx58NXvfKz2szefZXutgTSbaDI_20070608.html?mod=rss_free

quakes associated with glacial retreat and post-glacial rebound occurred at the end of the Pleistocene ice age, he says.[150]

Of course prediction is a hazardous pastime and the complexity of the systems we are considering means it would be silly to make any pronouncements about the times to come. However, if increased earthquakes and volcanoes came at a time of climatic stress it would serve to further disrupt ecosystems, agriculture, and weather in general.

Possibilities like this are not factored into current climate models, nor is it taken account in any of the optimistic scenarios suggesting how the environmental crisis can be dealt with. Most "solutions" are based upon simple linear models involving one, or at most a few, independent variables.

WILD CARD: METHANE HYDRATES

There is yet another feedback loop which may come into play. During the Permian-Triassic extinction about 251 million years ago about 90% of the earth's marine life and 70% of its land life forms went extinct suddenly. Massive volcanism in what is now Siberia appears to have triggered planetary warming. The effect of the sudden release of greenhouse gases by the volcanic event caused sufficient global warming to raise the temperatures of the oceans by 5 degrees Celsius. This in turn apparently triggered the rapid sublimation of undersea methane hydrate deposits, which raised global temperatures by an additional 5 degrees Celsius, or about 9 degrees Fahrenheit. This second very sudden temperature rise killed most of the living organisms on the planet. A similar process occurred during the Paleocene-Eocene thermal maximum about fifty-five and a half million years ago. [151]

Methane is about 20 times more powerful a heat-trapping gas than is carbon dioxide. Figure 10 below shows the worldwide distribution of known sub-sea methane hydrate deposits; the Indian Ocean and Atlantic African coast have not been adequately surveyed as of yet.

---

150. *Geotimes*, October, 2004, Melting Glaciers Promote Earthquakes, http://www.agiweb.org/geotimes/oct04/NN_glacier.html
151. Wikepedia The Free Online Encyclopedia: Permian-Triassic extinction event, http://en.wikipedia.org/wiki/Permian_extinction

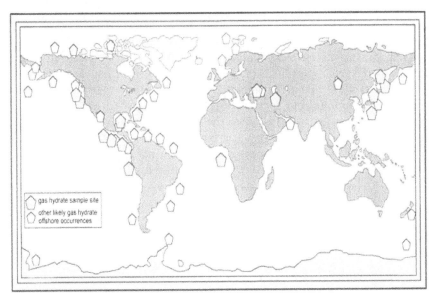

Figure 4. Worldwide Sub-Sea Distribution of Methane Hydrate Deposits.
Source: US Geological Survey, reproduced in Wikipedia.org.[152]

The oceans possess tremendous thermal inertia. However, once they begin warming, the process accelerates. Currently, "ocean surface temperatures worldwide have risen on average 0.9 degrees Fahrenheit, or 0.5 degrees Celsius, and ocean waters in many tropical regions have risen by almost 2 degrees F (1 degree C)...according to scientists."[153] It is conceivable that we will reach the triggering threshold for hydrate sublimation, setting off another chain reaction of warming.

So long as we look at each of the individual aspects of global warming in isolation from one another, nothing too dramatic seems to be happening. Atmospheric greenhouse gas concentrations in the atmosphere are increasing. Studies show oceanic currents are slowing in the North Atlantic, while the upwelling of deep, cold, nutrient rich, waters in the Pacific and elsewhere, is slowing, leading to a decrease in phytoplankton. But studies show that the earth has always moved through hotter and cooler periods — right? Right; only these dramatic shifts cause mass extinctions.

Corporate interests are profiting enormously from the industrial activities which are triggering all of these events and processes. In the aggregate even con-

152. US Geological Survey Image: Gas Hydrates, http://en.wikipedia.org/wiki/Image: Gas_hydrates_1996.jpg#filelinks
153. Oceans Alive, Wake-up Call: Oceans Warming Up, http://www.oceansalive.org/ explore.cfm?subnav=article&contentID=4704

cerned shareholders, when their interests are melded into the apersonal corporation, end up acting with a degree of amorality and self-centeredness because
the measures for success, for corporations, measure only money. Any corporation
that might actually seek to benefit mankind is liable to lose the competition for
money to a rival who stays "focused." And if you lose, you are out of the game.

Civilization's animating complex adaptive memetic system is unfit for the
purpose of maintaining our lives, because its fundamental organizing principle
of wealth creation is predicated upon the cancer-like rapid destruction of the
biological and natural systems upon which our own continued existence
depends. Yet this accelerating consumption of the natural world cannot continue. Either total collapse, or alternatively, planned fundamental change, must
ensue shortly.

Chapter 6. Famine, Disease, and Pestilence

If there be famine in the land, if there be pestilence, if there be blight, mildew, locust, caterpillar; if their enemy besiege them in the land of their gates; whatever plague, whatever sickness there be... — 1st Kings 8:37[154]

Disease and Pestilence

Again and again, the key insight of systems theory that "you can't do only one thing" keeps returning to our attention. As the world's climate warms — outside of Western Europe anyway — sudden ecological disruptions create new niches for opportunistic life forms such as insects and a variety of viral and microbial pathogens. Simultaneously with these developments, climate change seriously affects the procurement of plant and animal food resources. The results are famine, disease, pestilence, and inevitably, the deaths of humans on a mass scale.

Nowhere on the planet has experienced greater temperature increases over a short period of time than has the arctic region. According to a recently released study,[155] temperatures in Alaska have already warmed by 3.6 to 5.4 degrees during recent decades. One consequence of this warming was wholly unanticipated:

A massive beetle infestation has swept through millions of acres (hectares) in south-central Alaska over the past decade, scientists said, because significantly warmer weather is delaying the usual winter die-off of insect populations. The insects' voracious attack on spruce bark has left forests tinder-dry while general heat-induced stress have weakened forests, with lightning strikes making them a

---

154. *Online Parallel Bible*, 1 Kings 8:37, http://Bible.cc/1_kings/8-37.htm
155. Arctic Climate Impact Assessment, http://www.acia.uaf.edu/

fire hazard in the Chugach Mountain foothills, said Glenn Juday, a professor of forest ecology at the University of Alaska Fairbanks.[156]

Insects proliferate when temperatures rise. Many species of insects which carry human diseases, such as mosquitoes, are able to migrate to latitudes ever farther northwards and southwards, thereby bringing diseases such as West Nile fever into previously unaffected areas.

> "The common theme between diseases like malaria, yellow fever, dengue fever, St. Louis encephalitis and West Nile is an animal vector (through which they are spread), in this case the mosquito," said Dr. William H. Schlesinger, dean of the Nicholas School of the Environment and Earth Sciences and professor of biogeochemistry at Duke University. "Mosquitoes need wet places and a relatively warm climate to complete their life cycle. Any degree of global warming that extends the area of the globe conducive to that life cycle increases the potential for the rapid spread of these kinds of viruses." Dr. Paul R. Epstein, associate director of the Harvard Medical School Center for Health and the Global Environment, is even more direct: "We have good evidence that the conditions that amplify the life cycle of the disease are ... long-term extreme weather phenomena associated with climate change," he said.[157]

An international study co-sponsored by Harvard University's Center for Health and the Global Environment, Swiss Re (a reinsurance company) and the UN Development Program found that:

1. Warming favors the spread of disease.

2. Extreme weather events create conditions conducive to disease outbreaks.

3. Climate change and infectious diseases threaten wildlife, livestock, agriculture, forests and marine life, which provide us with essential resources and constitute our life-support systems.

4. Climate instability and the spread of diseases are not good for business.

5. The impacts of climate change could increase incrementally over decades.

6. Some impacts of warming and greater weather volatility could occur suddenly and become widespread.

7. Coastal human communities, coral reefs, and forests are particularly vulnerable to warming and disease, especially as the return time between extremes shortens.[158]

---

156. Planet Ark, Sept. 29, 2005, Alaska Landscape Transformed by Warmer Climate, http://www.planetark.com/dailynewsstory.cfm/newsid/32717/story.htm

157. Wired News, Oct. 1, 2002, Warmth May Feed West Nile Virus, http://www.wired.com/news/medtech/0,1286,55484,00.html

158. Climate Change Futures, Health, Ecological, and Economic Dimensions, A Project of The Center for Health and the Global Environment, Harvard Medical School, http://www.climatechangefutures.org/pdf/CCF_Report_Final_10.27.pdf

Across the planet, the reports are multiplying: Global warming is bringing insect infestations to areas such as Alaska which had previously been free of such plagues. Insects are spreading into new areas, bringing with them a host of deadly infectious diseases. Perversely, increased temperatures allow the insects to become more robust, thus allowing them to carry even greater viral loads:

> It's plain that global warming will have a major impact on human health, and that new threats will emerge. "With vector-borne diseases, it's pretty well established that increased temperatures cause higher and higher loads of viruses," says Peter Daszak, executive director of the Consortium for Conservation Medicine. "You also get an increase in biting rates of carriers like mosquitoes."[159]

Yet another consequence of global warming and the climatic disruptions it causes is a decline in insect predators, leading to still further increases in insect populations:

> Another prominent proponent of the West Nile global warming connection is Dr. Paul Epstein of Harvard University. "Droughts are more common and prolonged as the planet warms," he says. "Warm winters intensify drought because there's a reduced spring runoff. The cycle seems to rev up in the spring, as catch basin water dries up and what is left becomes organically rich and a perfect mosquito-breeding place. The drought also reduces populations of mosquito predators."[160]

Only a small percentage of the world's insect species, along with an even smaller percentage of its pathogens, have been cataloged and classified. As a result it should be no surprise that the World Health Organization has reported that at least thirty previously completely unknown human infectious diseases have emerged during the past 25 years.[161]

Additionally, rapid warming can release ancient viruses which have been locked away in ice for ages:

> Some scientists believe that climate change could unleash ancient illnesses as ice sheets drip away and bacteria and viruses defrost. Illnesses we thought we had eradicated, like polio, could reappear, while common viruses like human influenza could have a devastating effect if melting glaciers release a bygone strain to which we have no resistance. What is more, new species unknown to science may re-emerge. And it is not just humans who are at risk: animals, plants and marine creatures could also suffer as ancient microbes thaw out.[162]

This process of new diseases emerging and proliferating is only beginning. The reasons for this are because natural ecological constraints have been dis-

---

159. *E Magazine.com*, Nov/Dec, 2004, Too Darn Hot: Global Warming Accelerates the Spread of Disease, Vol. XV, Number 6, http://www.emagazine.com/view/?2116
160. Ibid., *E Magazine.com*, # 160.
161. *People and Planet.net*, 15 Mar. 2003, The threatened plague, http://www.peopleandplanet.net/doc.php?id=104
162. *The Independent*, 28 Sept. 2005, Global Warming: Death in the deep freeze, http://news.independent.co.uk/world/science_technology/article315614.ece

rupted by global warming. This gives potentially prolific opportunistic plant and animal species including pathogens, insects, along with rodents and weeds an opportunity for unchecked growth:

> Weeds, rodents, insects and micro-organisms are opportunists: They reproduce rapidly, have huge broods, small body sizes, wide-ranging appetites, and are good at dispersal and colonization of new environments. In stable environments large predators fare well and keep opportunistic species under control. But in degraded environments, opportunists can seize the upper hand; just as opportunistic infections take advantage in patients with weakened immune systems.[163]

What about their predators? Normally such opportunists would be severely constrained by a variety of predators and other environmental constraints. However global warming thoroughly disrupts the normal, healthy ecosystems within which such constraints operate:

> Several aspects of global change tend to reduce predators disproportionately, releasing prey from their biological controls. Among the most widespread of these are: habitat loss and fragmentation; monocultures in agriculture and aquaculture; excessive use of toxic chemicals; excess ultraviolet radiation; and climate change and weather instability.[164]

A study conducted by researchers from Duke and Harvard Universities was published in the Proceedings of the National Academy of Sciences in late May 2006. As reported by CBS News, the study's findings stated that:

> Compared to poison ivy grown in usual atmospheric conditions, those exposed to the extra-high carbon dioxide grew about three times larger — and produced more allergenic form of urushiol, scientists from Duke and Harvard University reported. Their study appears in this week's Proceedings of the National Academy of Sciences. "The fertilization effect of rising CO_2 on poison ivy ... and the shift toward a more allergenic form of urushiol have important implications for the future health of both humans and forests," the study concludes.[165]

As I've explained, disruption of the planet's climatic equilibrium will favor organisms which we consider to be pestilential — such as poison ivy. Conversely, the monocrops developed by industrial agribusiness, which require constant tending in conjunction with substantial inputs of fertilizers, pesticides, and mechanical harvesting, will almost certainly fail to thrive.

---

163. Ibid., People and planet.net, # 161
164. Ibid.
165. *CBS News*, Sci-Tech, Study: Global Warming Boosts Poison Ivy, May, 29, 2006, http://www.cbsnews.com/stories/2006/05/29/ap/tech/mainD8HTM3I00.shtml

OCEANIC DISRUPTION

Such proliferation of life forms that are harmful to mankind is not limited to the land. Algae and plankton blooms such as the red tide are damaging sea life. Marine mammals and sea birds are declining in numbers at alarming rates.[166] Coral reefs are dying while increased atmospheric CO_2 content has led to a commensurate increase in CO_2 dissolved in ocean waters. This has caused further consequences:

> "Terapods — free-swimming marine mollusks — are currently reacting to the changed conditions by dissolving their shells," Feely said. "What we do know for sure is the calcification rate will be reduced by 15 to 45% as a result of acidification," he said. These changes will have profound impacts on the health of coral reefs. By 2060, if present trends continue, "tropical corals will all be in a marginal state," he said. "There is no evidence that corals will be able to adapt to these changed conditions."[167]

Minke whale populations are now in freefall as well:

> Global warming has caused an unexpected collapse in the numbers of the world's most hunted whale, scientists believe. They think that a sharp contraction in sea ice in the Antarctic is the likeliest explanation behind new findings, which suggest that the number of minke whales in the surrounding seas has fallen by half in less than a decade.[168]

Deep sea stocks of fish including species such as tuna and orange roughy are being rapidly pushed towards extinction. Illegal fishing — catching fish above the limits set by the United Nations Fish Stocks Agreement — and other formal and informal fisheries agreements is one significant factor causing this decline.

An even more significant factor is bottom trawling. This practice amounts to strip mining the ocean down to actually scraping the seabed itself clean of all life, using huge weighted nets which scour the sea bottom on rollers. They can be as much as 60 meters (about 190 feet) wide and 10 to 15 meters (32 to 48 feet) high. Everything and anything in their path is scooped up. Coral and any and all marine organisms or even entire marine ecosystems in the nets path are annihilated. This is a cumulative, ongoing process involving numerous trawlers — and once thriving marine ecosystems are becoming marine deserts.

---

166. *The Independent*, 13 Nov., 2005, Fish numbers plummet in warming Pacific, http://news.independent.co.uk/environment/article326752.ece

167. *Chemical and Engineering News*, March 16, 2005, Stark Effects From Global Warming: CO_2 emissions are causing oceans to warm, ocean chemistry to change, and rainfall patterns to shift, http://pubs.acs.org/cen/news/83/i12/8312globalwarming.html

168. *BBC News International*, Deep Sea Fish Stocks 'Plundered', 19 May, 2006, http://news.bbc.co.uk/2/hi/science/nature/4996268.stm#trawl

BBC News, commenting on a study by the World Wildlife Foundation (WWF) on these issues, states:

> Fish stocks in international waters are being plundered to the point of extinction, a leading conservationist group has said. Illegal fishing and bottom-trawling in deep waters are to blame, according to a report from WWF. It says the current system of regional fishing regulation is failing to tackle the problem, with not enough being done to enforce quotas or replenish stocks. It says species under severe threat include tuna and the orange roughy. The orange roughy is targeted by bottom-trawlers, which drag heavy rollers over the ocean floor, destroying coral and other ecosystems. "Given the perilous overall state of marine fisheries resources and the continuing threats posed to the marine environment from over-fishing and damaging fishing activity, the need for action is immediate," Simon Cripps, director of WWF's global marine programme, said. Illegal fishing "by highly mobile fleets under the control of multinational companies" was identified as one of the worst threats to marine life. But the report also attacked governments for over fishing. "Vast over-capacity in authorised fleets, over-fishing of stocks... the virtual absence of robust rebuilding strategies... and a lack of precaution where information is lacking or uncertain are all characteristic of the management regimes currently in place," it said.[169]

The study's executive summary notes that at the present time nearly all of the planet's fisheries are managed by Regional Fisheries Management Organizations, which are commonly called RFMOs. These regional organizations are a type of international regime which deal with a specific issue or related set of issues. In this case, they are focused narrowly upon fisheries management. RFMOs represent agreements between states which regulate fishing in particular areas of the oceans for certain species of fish, to apportion the catches in such a manner as to not over-harvest the stocks of fish, in order to prevent a population crash. These regional agreements are coordinated globally via the United Nations Fish Stocks Agreement, the UNFSA. The World Wild Life Foundation's report finds that:

> Despite the proliferation of RFMOs and the development and evolution of instruments aimed at empowering them RFMOs have generally failed to prevent over-exploitation of straddling and highly migratory fish stocks, to rebuild overexploited stocks and to prevent degradation of the marine ecosystems in which fishing occurs. Not only have broader, international expectations not been met but RFMOs have also largely failed to meet the objectives of their own governing conventions, generally characterized as conservation and sustainable utilization of target stocks under their mandate. It is difficult to identify examples of sustainable management of target stocks by RFMOs.[170]

---

169. Willock, A., and Lack, M., 2006, Follow the leader: Learning from experience and best practice in regional fisheries management organizations, WWF International and TRAFFIC International, http://news.bbc.co.uk/1/shared/bsp/hi/pdfs/19_05_06_wwf_report.pdf

The WWF study is predicated upon the assumption that all that is required to resolve these over fishing problems is better coordination through the UN and more efficient enforcement action by its member states. The unspoken assumption is that the political economy of the planet is fundamentally sound, as is its overall ecology. If both of these assumptions are incorrect, it follows that the proposed remedies have little chance of success.

Meanwhile, other climatic effects will disrupt the world's marine ecosystems. If there are major shifts in the ocean's currents, it will have massive effects upon marine life as well.

> Circulation patterns also deliver nutrient-rich waters to strategic parts of the ocean. A disruption of the ocean conveyor would interfere with this delivery system of nutrient supplies to sea animals and could have dire consequences for the marine web of life. As surface waters heat up, the vertical layers of sea water could mix less with each other, an effect called vertical stratification. Upwellings of cold, nutrient-rich waters would become less frequent, diminishing blooms of phytoplankton, microscopic plants that anchor the marine food chain. On top of that, phytoplankton use carbon dioxide for photosynthesis. If plankton become depleted, the oceans could not remove as much carbon dioxide from the atmosphere. The marine food chain may already be showing signs of breaking. This year on the US West Coast and last year in Britain, hundreds of thousands of seabirds failed to breed. Dead seabirds like cormorants and Cassin's auklets have washed up on West Coast beaches. Juvenile rockfish counts are the lowest they've been off California in more than 20 years. Most alarming, small crustaceans like krill — the base of the ocean's food web — have suffered steep declines. The culprit for the collapse appears to be slackening upwellings, which have decreased phytoplankton blooms in these coastal areas. Fewer phytoplankton mean fewer fish, leaving the birds to face mass starvation. Scientists speculate that this decrease in their food supply could be an effect of global warming.[171]

This is consistent with Prof. Huisman's research, presented in the previous chapter, concerning the adverse effects of global warning on phytoplankton. These phytoplankton form the foundation for the entire global oceanic ecosystem and marine food chain.

Phytoplankton also produce at least 50% of the planet's oxygen.[172] Tropical rainforests, occupying six percent of the world's land surface, produce about 40 percent.[173] The Rainforest Action Network states that: "If defores-

---

170. *National Geographic News*, Aug. 1, 2001, Whale Population Devastated by Warming Oceans, Scientists Say, http://news.nationalgeographic.com/news/2001/08/0801_wirewhales2.html

171. Oceans Alive, Glaciers and the Food Chain, http://www.oceansalive.org/explore.cfm?subnav=article&contentID=4706

172. *National Geographic News*, June 7, 2004, Source of Half the Earth's Oxygen Gets Little Credit, http://news.nationalgeographic.com/news/2004/06/0607_040607_phytoplankton.html

tation continues at current rates, scientists estimate nearly all tropical rainforest ecosystems will be destroyed by the year 2030."[174]

Oceanic phytoplankton are in rapid decline, and so are the tropical rainforests which account for most of the other half. It will be difficult to find a substitute for oxygen....

FAMINE

Scientists from the University of Wisconsin-Madison combined agricultural data with global satellite images to analyze the ways in which the earth's surface was being used. They found that about 40% of the land surface was presently being used for agriculture, up from an estimated 7% percent in 1700. They found that:

> ....the earth is rapidly running out of fertile land and that food production will soon be unable to keep up with the world's burgeoning population....The Amazon basin has seen some of the greatest changes in recent times, with huge swaths of the rainforest being felled to grow soya beans. "One of the major changes we see is the fast expansion of soybeans in Brazil and Argentina, grown for export to China and the EU," said Dr Ramankutty. This agricultural expansion has come at the expense of tropical forests in both countries. Meanwhile, intensive farming practices mean that cropland areas have decreased slightly in the US and Europe and the land is being gobbled up by urbanisation. The research indicates that there is now little room for further agricultural expansion. "Except for Latin America and Africa, all the places in the world where we could grow crops are already being cultivated. The remaining places are either too cold or too dry to grow crops," said Dr Ramankutty.[175]

Logically, it seems the world's ability to feed a growing human population must eventually reach a limit. There are some indications that the limit is closer than almost anyone realizes:

> The world is now eating more food than farmers grow, pushing global grain stocks to their lowest level in 30 years. Rising population, water shortages, climate change, and the growing costs of fossil fuel-based fertilisers point to a calamitous shortfall in the world's grain supplies in the near future, according to Canada's National Farmers Union (NFU). Thirty years ago, the oceans were teeming with fish, but today more people rely on farmers to produce their food than ever before, says Stewart Wells, NFU's president. In five of the last six years, global population ate

173. Blue Planet Biomes: Tropical Rainforest, http://www.blueplanetbiomes.org/rainforest.htm; *Wikipedia, The Free Online Encyclopedia*: Rainforest, http://www.reference.com/browse/wiki/Rainforest

174. Rainforest Action Network, Rates of Rainforest Loss: Rainforest fact sheets, http://www.ran.org/info_center/factsheets/04b.html

175. *Guardian Unlimited*, Dec. 6, 2005, Food Crisis Feared as Fertile Land Runs Out, http://www.guardian.co.uk/food/Story/0,,1659112,00.html

significantly more grains than farmers produced. And with the world's farmers unable to increase food production, policymakers must address the "massive challenges to the ability of humanity to continue to feed its growing numbers", Wells said in a statement. There isn't much land left on the planet that can be converted into new food-producing areas, notes Lester Brown, president of the earth Policy Institute, a Washington-based non-governmental organisation. And what is left is of generally poor quality or likely to turn into dust bowls if heavily exploited, Brown told IPS.[176]

It is at this very moment that the nitrogen fertilizers which are derived from natural gas are about to become more expensive and hard to come by. The same is true for the gasoline required to power agricultural machinery and to process and ship agricultural products across the world. The world's ability to feed itself is facing enormous challenges, as the human population rises. The following report was released in 2006:

> This year's world grain harvest is projected to fall short of consumption by 61 million tons, marking the sixth time in the last seven years that production has failed to satisfy demand. As a result of these shortfalls, world carryover stocks at the end of this crop year are projected to drop to 57 days of consumption, the shortest buffer since the 56-day-low in 1972 that triggered a doubling of grain prices. World carryover stocks of grain, the amount in the bin when the next harvest begins, are the most basic measure of food security. Whenever stocks drop below 60 days of consumption, prices begin to rise. It thus came as no surprise when the US Department of Agriculture (USDA) projected in its June 9 world crop report that this year's wheat prices will be up by 14 percent and corn prices up by 22 percent over last year's. This price projection assumes normal weather during the summer growing season. If the weather this year is unusually good, then the price rises may be less than those projected, but if this year's harvest is sharply reduced by heat or drought, they could far exceed the projected rises. With carryover stocks of grain at the lowest level in 34 years, the world may soon be facing high grain and oil prices at the same time. For the scores of low-income countries that import both oil and grain, this prospect is a sobering one.[177]

This appears to be a long-term trend and not just normal supply and demand, good year/bad year climate fluctuations about an optimal mean. The shortages are intrinsic to the system of industrial agriculture itself. Peak oil means peak food production too!

In addition, the situation is exacerbated as more and more of the world follows the trend of urbanization. Formerly self-sufficient agrarian populations are driven to seek sweatshop labor jobs in the slums of cities, dislocated by the spread of industrial factory farming in conjunction with cheap imports of corporately farmed food from wealthy countries.

---

176. *Inter Press News Service Agency*, May 17, 2006, Population: Global Food Supply Near the Breaking Point, http://www.ipsnews.net/news.asp?idnews=33268
177. Earth Policy Institute, June 15, 2006, World Grain Stocks Fall to 57 Days of Consumption, Grain Prices Starting to Rise, http://www.earth-policy.org/Indicators/Grain/2006.htm

	Percentage urban			
	1950	*1975*	*2000*	*2030*
World	29.7	37.9	47.0	60.3
More developed regions	54.9	70.0	76.0	83.5
Less developed regions	17.8	26.8	39.9	56.2

Table 1. World Urbanization for More and Less Developed Regions: 1950-2030.
Source: United Nations World Urbanization Prospects 1999 Revision.[178]

	Percentage urban			
	1950	*1975*	*2000*	*2030*
Northern America	64	74	77	84
Latin America and the Caribbean	41	61	75	83
Europe	52	67	75	83
Oceania	62	72	70	74
Africa	15	25	38	55
Asia	17	25	37	53

Table 2. World Urbanization by Continent: 1950-2030.
Source: United Nations World Urbanization Prospects 1999 Revision.[179]

This movement to greater urbanization comes at a time when the planet's ecological systems are approaching anthropogenically-caused collapse; and when disruptions to its hydrological systems — consistent rainfall patterns and life giving oceanic currents — are reaching a critical threshold.

As shown by the above Tables 1 and 2, and by Figure 1, the United Nations data show that the spread of industrialization across the entire planet is rapidly shifting the majority of inhabitants of the less industrialized "third world" nations off of the land and into densely crowed urban environments. These UN figures indicate that within the next few years, the majority of humanity will be living in urban or suburban environments. Specifically, in its 2003 revision of its *World Urbanization Prospects* study, the UN researchers concluded that:

> While 48 per cent of the world's population is estimated to live in urban areas in 2003, current projections indicate that the fifty per cent mark will be crossed in

---

178. United Nations, World Urbanization Prospects, the 1999 Revision: Key Findings, http://www.un.org/esa/population/pubsarchive/urbanization/urbanization.pdf
179. Ibid., United Nations, 1999 Revision, # 178

2007; thus, for the first time in history the world will have more urban dwellers than rural ones. The proportion of the population that is urban is expected to rise to 61 per cent by 2030.[180]

The UN's 2003 study indicated that the rate of urbanization in the third world was increasing during the 1999 to 2003 study period. Following this, in 2005 the United Nations Food and Agriculture Organization (FAO) issued a report which stated that:

> ....scientific studies show that global warming would lead to an 11 per cent decrease in rain-fed land in developing countries and in turn a serious decline in cereal production. "Sixty-five developing countries, representing more than half of the developing world's total population in 1995, will lose about 280 million tons of potential cereal production as a result of climate change," FAO said. The effect of climate change on agriculture could increase the number of people at risk of hunger, particularly in countries already saddled with low economic growth and high malnourishment levels. "In some 40 poor, developing countries, with a combined population of 2 billion... production losses due to climate change may drastically increase the number of undernourished people, severely hindering progress in combating poverty and food insecurity," the report said.[181]

In other words, just as the majority of humanity becomes dependent upon industrialized agriculture for their food supply, the stable weather patterns required to produce the food are disappearing. This occurs simultaneously with the rapid spread of opportunistic organisms such as microbial pathogens, rodents, weeds, and so on. And the cause of these several crises is identical: human-caused disruption of historically stable ecological interrelationships.

There is yet another dimension to the provision of food for a globally urbanized population: industrial agriculture is completely dependent upon hydrocarbon inputs.

Natural gas is the primary material used in manufacturing fertilizer, and vast amounts of natural gas are utilized to restore nitrogen-depleted soils annually. Pesticides, manufactured from petroleum, are used continuously and massively in industrial agriculture. Additionally, large quantities of hydrocarbons are utilized for plowing, reaping, and processing agricultural products. These products are then transported in hydrocarbon-burning vehicles, for further hydrocarbon-powered processing, packaging and transportation to their points of sale across the planet.

Indeed, industrial agriculture's dependence upon petroleum is so overwhelming that we can very literally be said to be eating oil and natural gas with every meal we consume. In a meticulous analysis of the energy implications of industrial agriculture, Richard Manning found that:

---

180. United Nations, World Urbanization Prospects, the 2003 Revision, http://www.un.org/esa/population/publications/wup2003/WUP2003Report.pdf

181. *ABC News Online*, May 27, 2005, Climate Change Likely to Increase Famine: FAO, http://www.abc.net.au/news/newsitems/200505/s1378213.htm?news_id=16

The common assumption these days is that we muster our weapons to secure oil, not food. There's a little joke in this. Ever since we ran out of arable land, food is oil. Every single calorie we eat is backed by at least a calorie of oil, more like ten. In 1940 the average farm in the United States produced 2.3 calories of food energy for every calorie of fossil energy it used. By 1974 (the last year in which anyone looked closely at this issue), that ratio was 1:1. And this understates the problem, because at the same time that there is more oil in our food there is less oil in our oil. A couple of generations ago we spent a lot less energy drilling, pumping, and distributing than we do now. In the 1940s we got about 100 barrels of oil back for every barrel of oil we spent getting it. Today each barrel invested in the process returns only ten, a calculation that no doubt fails to include the fuel burned by the Hummers and Blackhawks we use to maintain access to the oil in Iraq. David Pimentel, an expert on food and energy at Cornell University, has estimated that if all of the world ate the way the United States eats, humanity would exhaust all known global fossil-fuel reserves in just over seven years. Pimentel has his detractors. Some have accused him of being off on other calculations by as much as 30 percent. Fine. Make it ten years.[182]

"Every single calorie we eat is backed by at least a calorie of oil, more like ten." Notice that the ratio of food-calories consumed to oil-calories expended to grow the food is not fixed. Rather, oil and natural gas will become progressively more difficult to extract and refine, because all of the easy-to-reach-and-refine, "low hanging fruit," pools of petroleum have been used already. More and more energy will have to be used to obtain less and less new hydrocarbon energy.

In tandem with this, the increase in urbanization worldwide means that more and more of the planet's agriculture becomes industrial-scale and requires massive hydrocarbon inputs, unlike traditional peasant agriculture. At this very moment of agricultural transition, ecological disruption and increasing pestilence will rapidly decrease crop yields.

The result is that ever greater quantities of ever more expensive oil, obtained at an ever decreasing EROEI (Chapter 3) must be used to keep alive an ever growing human population which is ever more urban and hence ever more dependent upon industrial agriculture. All of this at the exact time in history when supplies of hydrocarbon energy sources are commencing their inexorable decline.

Inevitably, food production per capita must decline precipitously. Food production cannot increase significantly by means of placing additional lands into cultivation, because, as we have already seen, there is not much more arable land available.

Industrial agriculture is absolutely dependant upon the input of vast quantities of oil and natural gas. These quantities must begin to decrease in absolute terms even as population continues to increase. Consider projected human population growth over the next decade:

---

182. *Harpers.org*, July 23, 2004, The Oil We Eat, http://www.harpers.org/TheOil-WeEat.html

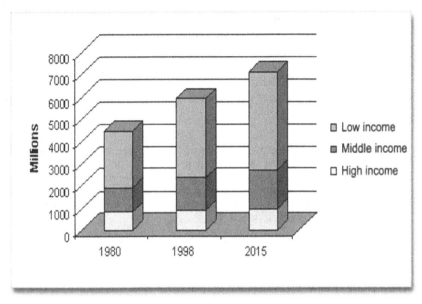

Figure 1. Population Increase by Income Category: 1980-2015.

Source: World Bank.[183]

The high and middle income populations which are already almost completely urbanized grow slowly (middle income) or not at all (high income). However, the low income population of the planet will continue to grow rapidly. At the same time, there is an even more rapid trend of low-income and self-sufficient peasant farmers leaving the land and becoming urban residents, who are wholly dependent upon industrial agriculture. The UN's World Urbanization Prospects 2003 Revision found that:

> During 2000-2030, the world's urban population will grow at an average annual rate of 1.83 per cent, nearly double the rate expected for the total population of the world (nearly 1 per cent per year). At that rate of growth, the world's urban population will double in 38 years. Growth will be particularly rapid in the urban areas of less developed regions, averaging 2.29 per cent per year during 2000-2030, consistent with a doubling time of 30 years. In contrast, the rural population of the less developed regions is expected to grow very slowly, at just 0.1 per cent per year during 2000-2030.[184]

---

183. The World Bank, Development Education Program, Population Growth Rate, http://www.worldbank.org/depweb/english/modules/social/pgr/chart1.html
184. People and Planet.net, Urban Population Trends, http://www.peopleandplanet.net/doc.php?id=1489

Consider all of the factors involved here: the population increases, but the land available for cultivation cannot be increased. Hydrocarbon inputs must decrease rapidly. Climate becomes less optimal, while pestilence increases.

This bleak conclusion is further reinforced by recent findings reported by the Chinese Academy of Sciences. This study finds that warming throughout the Tibetan plateau is rapidly melting its remaining 60,000 square miles of glaciers. The Chinese research estimates that the area covered by glaciers will decrease by about half every decade.

> The melting threatens to disrupt water supplies over much of Asia. Many of the continent's greatest rivers — including the Yangtze, the Indus, the Ganges, the Brahmaputra, the Mekong, and the Yellow River, rise on the plateau. In China alone, 300 million people depend upon water from the glaciers for their survival. Yet the plateau is drying up, threatening to escalate an already dire situation across the country. Already 400 cities are short of water; in 100 of them — including Beijing — the shortages are becoming critical.[185]

The lives of more than a billion people are imperiled by this development.

This process of mid-latitude glacial melt is occurring with ever increasing rapidity across the planet. For example a team of British researchers found that all glaciers in Africa would vanish within two decades:

> "Recession of these tropical glaciers sends an unambiguous message of a changing climate in this region of the tropics," said lead researcher Richard Taylor of the University College of London, Department of Geography. A century ago the Rwenzori glaciers were surveyed at 2.5 square miles. The area covered by glaciers halved between 1987 and 2003 and is now down to about 0.4 square mile, the researchers said. They said the glaciers are expected to disappear within the next 20 years if present trends continue.[186]

This development exacerbates another one. As reported by the BBC a just released report by the International Center for Soil Fertility and Agricultural Development (IFDC) has found that:

> Africa's farmland is rapidly becoming barren and incapable of sustaining the continent's already hungry population, according to a report. The report shows that more than 80% of the farmland in Sub-Saharan Africa is plagued by severe degradation. This is a major cause of poverty and hunger in sub-Saharan Africa, where one in three people is undernourished...More than 60% of Africa's population is directly engaged in agriculture. But crop productivity has remained stagnant, while cereal

185. *The Independent, Online Edition*, May 7, 2006, Ice-Capped Roof of the World Turns to Desert, Scientists warn of ecological catastrophe across Asia as glaciers melt and continent's great rivers dry up, http://news.independent.co.uk/environment/article362549.ece

186. *Environmental News Network*, May, 16, 2006, Glaciers in Africa Expected to Disappear, http://www.enn.com/today.html?id=10463

BBC News International, March 30, 2006, "Barren Future" for Africa's Soil, http://news.bbc.co.uk/1/hi/sci/tech/4860694.stm

yields in Asia have risen three-fold over the past four decades....During the 2002-2004 cropping season, about 85% of African farmland had nutrient depletion rates of more than 30kg per hectare yearly. About 40% of farmland had nutrient depletion rates greater than 60kg per hectare yearly. In addition to removal by crop harvests, other factors contributing to nutrient depletion include loss of nitrogen and phosphorus through soil erosion by wind and water, and leaching of nitrogen and potassium.[187]

Rapid and destabilizing climate change, including desertification and aridification, as indicated by the disappearance of its glaciers, is yet another factor which will affect African agricultural production negatively which the above study does not take into account. Further, existential crisis across the planet will essentially eliminate the ability of the developed world to assist in any meaningful way. Outside of Asia and Africa the picture is equally bleak:

> By the century's end, the Andes in South America will have less than half their current winter snowpack, mountain ranges in Europe and the US West will have lost nearly half of their snow-bound water, and snow on New Zealand's picturesque snowcapped peaks will all but have vanished. Such is the dramatic forecast from a new, full-century model that offers detail its authors call "an unprecedented picture of climate change." The decline in winter snowpack means less spring and summer runoff from snowmelt. That translates to unprecedented pressure on people worldwide who depend on summertime melting of the winter snowpack for irrigation and drinking water. Hardest hit are mountains in temperate zones where temperatures remain freezing only at increasingly higher elevations, said Steven J. Ghan, staff scientist at the Department of Energy's Pacific Northwest National Laboratory and lead author of a study describing the model in the current Journal of Climate. PNNL scientist Timothy Shippert was co-author.[188]

Across the planet, water supplies for drinking, for crop irrigation and for industry, will become increasingly constrained. Already half of the world's major rivers have become either depleted or polluted, or both.[189] At a time when demand upon water supplies for each of these purposes will be rapidly rising, actual water supplies will be rapidly falling. The consequences of this are many and grim.

In some cases these consequences are less than obvious. Each time a "solution" is proposed to any one crisis facing humanity, there is an implicit

---

187. *BBC News International*, March 30, 2006, 'Barren Future' for Africa's Soil, http://news.bbc.co.uk/1/hi/sci/tech/4860694.stm

188. EurekAlert!, 18 May, 2006, New century of thirst for world's mountains, Most detailed forecast to date shown sharp snowpack decline between now and year 2100; New Zealand, Latin America, Western US, European ranges hardest hit, http://www.eurekalert.org/pub_releases/2006-05/dnnl-nco051806.php

189. *The Independent, Online Edition*, June 7, 2006, Death of the world's rivers, disaster warning from UN as investigation reveals half of the planet's 500 biggest rivers are seriously depleted or polluted, http://news.independent.co.uk/environment/article350785.ece

assumption that all else will remain unchanged. For example, development of unconventional hydrocarbon resources such as the tar sands in Alberta, Canada require the application of vast quantities of water, which becomes polluted and unusable. At present, water is not seen as being a constraint to scaling up the production of this resource. However, it will be in a decade's time, and more so in twenty years' time.

And so for yet another unacknowledged reason, optimistic projections of solutions to one set of problems or another fail due to the intervening effects of all of the other sets of problems. Worse, even bleaker assessments generally assume rational decision-making by humanity's political authorities. As we shall see, that will not be forthcoming.

In early March of 2006 Britain's Defense Secretary, John Reid issued a strong warning:

> ...global climate change and dwindling natural resources are combining to increase the likelihood of violent conflict over land, water and energy. Climate change, he indicated, "will make scarce resources, clean water, viable agricultural land even scarcer" — and this will "make the emergence of violent conflict more rather than less likely." Although not unprecedented, Reid's prediction of an upsurge in resource conflict is significant both because of his senior rank and the vehemence of his remarks. "The blunt truth is that the lack of water and agricultural land is a significant contributory factor to the tragic conflict we see unfolding in Darfur," he declared. "We should see this as a warning sign."[190]

We must understand that climate change as it accelerates will not be a gradual and incremental process. Instead it will involve sudden changes in climatic conditions. Drought, flooding, and severe storms are probable outcomes.

The entire global political economy hinges upon an assumption that climate will not change significantly. And this assumption is in the process of being violated.

The Earth's oceans cover 70% of the planet's surface. Because water is denser than air, it stores about four times more heat per unit of volume than air does. Taking the vastly greater mass of oceanic water compared to atmospheric air, the world's oceans store about one thousand times more heat than does the atmosphere. They thus act, initially, to slow the process of global warming. But only for a time, and then climate can become unstable across the globe as the planet is knocked out of what scientists call "energy balance."

When this occurs the stable, predictable, and consistent weather needed for agriculture to work will no longer be the rule.

Civilization is fragile. It represents a slowly constructed complex-adaptive system composed of memes, which regulate the interactions of humans with

---

190. *Tom Paine.com*, March 7, 2006, The Coming Resource Wars, http://www.tompaine.com/articles/2006/03/07/the_coming_resource_wars.php

each other and with the surrounding biological and natural worlds. If we select unfit memes for these purposes — and we have indeed done so — then the patient efforts of millennia can be undone in very short order.

CHAPTER 7. POLITICAL FAILURE, AND ECONOMIC COL-LAPSE

In a time of universal deceit, telling the truth becomes a revo-lutionary act. — George Orwell[191]

POLITICAL FAILURE

Civilization has been organized on the premise that the world is a vast, essentially limitless storehouse of natural wealth; and on the related premise that those who are ruthless enough to defeat their rivals in accumulating the greatest pile of wealth are heroes and role models.

A system which functions solely in the quest for unending growth to produce greater material wealth — for a fortunate few — in the shortest pos-sible time, is not designed to allow for any questions of morality. That has been "designed out." Our core values placed wealth and comfort above all else.

Further, we have enshrined one method of organizing human abilities for this socially sanctioned purpose above all others. We have actually granted human rights to the artificial constructs that exist solely to gather wealth. By doing so, we have selected particular short-term material wealth producing memes, which have in turn set human civilization upon what would appear to be an irreversible trajectory towards several crises.

Of course, all major players in the oil business — private and public — insist that there will be enough oil to last well through the 21st century. But given their incen-tive to inflate reserve totals, it would be irresponsible not to question their esti-mates. The official figures — that is, those cited by oil companies to prove their product is secure — are notoriously unreliable. For example, in 2002 the US Geo-logical Survey claimed that total US oil production would eventually reach 362 bil-

---

191. Wisdom Quotes: George Orwell, http://www.wisdomquotes.com/002665.html

lion barrels. This calculation far surpasses most independent estimates, which place the figure closer to 200 billion; furthermore, it would require new American discoveries to equal the total reserves of Kuwait. The USGS itself admits its figures are "based on non-technical considerations that support domestic supply growth to the levels necessary to meet domestic demand levels." In other words, it determines supply estimates not by how much oil is left, but by guessing how much people will ultimately want.[192]

In plainer words, these are just made up numbers. The US Government projections of the future supply of oil are fabricated!

In today's world the principal centers of power are not the elected or unelected leaders of national governments but are in fact the multinational corporations. Of these, none are wealthier or more powerful than the global energy companies. It is in these energy corporations' interest to promulgate the fiction that hydrocarbon energy supplies are essentially unlimited; and this is exactly the message that is being disseminated in the media and the for-hire "think tanks" that give a semblance of objectivity to such claims of unlimited resources. To lend further credibility, the appropriate scientific civil service agency — the US Geological Survey — is pressed into service.

In principle, a corporation is a sort of tool to allow investors to contribute assets to some commonly agreed upon purpose. Their liability, in the eventuality that the corporation fails, is limited by law to only those assets the investors had contributed to the venture. In essence, a corporation is a legal fiction, designed to allow investors to receive potentially unlimited gains from their investments while limiting their potential losses to only those assets they have contributed to the corporation. The corporation thus becomes a legally defined entity, separate from the humans who have invested in it. This legal fiction can be quite useful: investors who wish to finance a start-up company effectively gamble that their investments will pay off in future corporate earnings. If the corporation instead goes bust, leaving huge liabilities, they and their personal assets are legally shielded from those liabilities. This allows for economic risk-taking which otherwise would not occur.

In an era in which the rapid growth of scientific knowledge can potentially be translated into new kinds of wealth creation, such legal incentives are arguably beneficial not only to the investors, but to society as well. New wealth and new techniques for doing things that may possibly contribute to human well-being are pioneered by economic risk-takers, who are generally called entrepreneurs.

---

192. *SFGate.com*, April 2, 2004, Fossil Fuel Dependency: Do Oil Reserves Foretell Bleak Future?, http://www.sfgate.com/cgi-bin/article.cgi?f=/c/a/2004/04/02/EDGGG5UJMG1.DTL&hw=Fossil+fuel+dependency&sn=032&sc=294; and DOE-EIA-0383(98), Dec. 1997, Annual Energy Outlook 1998 With Projections to 2020, http://tonto.eia.doe.gov/FTPROOT/forecasting/038398.pdf

The existence of corporations is thus not the actual problem. The core problem came about when we gave these artificial constructs the legal status of actual persons. The potential for this occurring, and its far reaching implications, were clearly foreseen by a few of the American Republic's founding fathers — Thomas Jefferson in particular.

JEFFERSON WAS RIGHT

Thomas Jefferson and James Madison tried hard to have an 11th Amendment included in the nation's original Bill of Rights. Their proposed Amendment would have prohibited "monopolies in commerce."[193] The amendment would have made it illegal for corporations to own other corporations, or to give money to politicians, or to otherwise try to influence elections. Corporations would have been chartered by the states for the primary purpose of "serving the public good."[194] Corporations would have possessed the legal status not of natural persons but rather of "artificial persons."[195] This means that they would have had only those legal attributes which the state saw fit to grant to them. They would not, and indeed could not, possess the same bundle of rights which actual flesh and blood persons enjoy, such as the right to free speech under the First Amendment. Under this proposed Amendment, neither the 14th Amendment of the US Constitution, nor any other provision for protection of personal rights and liberties that are included in the Constitution, would apply to those artificial entities called corporations.

Jefferson and Madison were insistent upon this amendment because the domination of colonial economic and political life by the greatest multinational corporation of its age — the British East India Company — was a driving factor behind the Revolution. It was the British East India Company that owned the tea that Sam Adams and his friends dumped overboard in Boston Harbor; it was the British East India Company that was responsible for the taxes on commodities and restrictions on trade that had been imposed on the colonists. In the end, a majority in the first Congress believed that already existing state laws governing corporations were adequate for constraining corporate power. Jefferson worried about the growing influence of corporate power until his dying day in

---

193. *The Founders Constitution*, Volume 4, Article 7 Document 12, Thomas Jefferson to Alexander Donald, 7 Feb., 1788, Papers 12:571, http://press-pubs.uchicago.edu/founders/ documents/a7s12.html; http://press-pubs.uchicago.edu/founders/documents/ v1ch2s23.html; and http://press-pubs.uchicago.edu/founders/documents/ v1ch2s23.html

194. Hartman, Thom, *Unequal Protection, The Rise of Corporate Dominance and the Theft of Human Rights*, St. Martin's Press, New York, NY, 2002., pps. 74-94, # 198

195. Ibid., Hartman, Thom, pps. 74-95, # 194

1826. The even more conservative founder, John Adams, also came to harbor deep misgivings about unchecked corporate power.

A few years after Jefferson's unsuccessful attempt to incorporate this amendment into the Bill of Rights, the fourth Chief Justice of the US Supreme Court, John Marshall, unilaterally asserted the Court's right to judicial review in the seminal case of *Marbury v. Madison*[196] in 1803. This decision meant that the Supreme Court would have sole and unchecked power to determine what the Constitution meant.

Jefferson was aghast. His fear lay in the knowledge that an unelected branch of government, one which is not subject to the will of the citizens, and is effectively immune from check by the two elected branches of government (only one Supreme Court Justice has ever been impeached — and none have ever been convicted and removed from office) was now solely responsible for determining the meaning of the Constitution. The meaning of the Constitution, and hence the very nature of our political system, was now in the hands of an un-elected and effectively uncontrollable body. "The Constitution has become a thing of wax to be molded as the Court sees fit," Jefferson lamented.[197] [198]

In 1886, Jefferson's two greatest Constitutional fears came true and effectively derailed true democracy in the United States; and other nations since have either copied the American example or have fallen under the dominion of multinational corporations — or both. The precipitating event was the case of *Santa Clara County v. Southern Pacific Railroad*.[199] This case is cited to this present day as having conferred the status of natural personhood, as opposed to artificial personhood, upon American corporations.

However, the Supreme Court had in fact actually declined to rule on the issue. J.C. Bancroft Davis, the Clerk of the Court, an attorney who curiously was also a former railroad company president, used his position to simply write this conclusion into the head notes which summarized the case. Due to Davis's falsification of the Supreme Court's decision and effective rewriting of the Constitution, corporations have had the status of actual persons whose rights are fully protected by the Constitution ever since.

This was a coup against democracy which succeeded because there were no real external checks and balances on the Court, and because the Court itself

196. Find law for legal professionals, US Supreme Court, Marbury vs. Madison, 5 U.S. 137, (1803).

197. Ibid., Hartman, Thom, p. 78, # 194.

198. *The Founders Constitution*, Volume 3, Article 1, Section 8, Clause 18, Document 16, Sept. 6, 1819, Works 12: 135-138, http://press-pubs.uchicago.edu/founders/documents/a1_8_18s16.html

199. Findlaw.com For legal professionals, Santa Clara County vs. Southern Pac. R. Co., 118, U.S. 394 (1896), http://caselaw.lp.findlaw.com/cgi-bin/getcase.pl?court=US&vol=118&invol=394

chose not to act to repudiate Davis's rewriting of the Constitution. The thing stood. Precedent was established. Jefferson's nightmare had come to pass.

And unlike a real person, a corporation is not even subject to death as a natural limit to its span of action and influence. Neither labor unions nor any other category of special interest group possess this attribute of "personhood" that has been granted to corporations, and so these interest groups are fundamentally and intrinsically unable to compete against corporate persons.

A seamless web of corporate power now connects multinational corporations with the mass media. This military-industrial-media complex controls a political candidate's access to publicity, and the positive or negative nature of that publicity. The military-industrial-media control access to money as well as determine how a candidate will be presented to the viewers. Therefore, the military-industrial-media complex largely determines which politicians will and will not get elected. And thus they control the government.

The policies that elected officials and candidates can espouse are rigorously circumscribed. With some notable exceptions, the corporate media has seen to it that the average American is woefully uninformed. The voter's primary role is simply to consume and to spend money. We are to be consumers and corporate subjects, not citizens. Under this materialistic system, live is devoid of deep meaning; we are conditioned to work ever harder and go ever deeper in debt to accumulate ever more useless junk, as though if we just piled up enough stuff we would somehow, magically, become happy.

The growing mismatch between our economy and our political system creates a national disconnect. Nations are geographically bounded, and their politics are localized; however, our economy is increasingly global in scope. Large corporations can treat the entire world as a seamless whole for business purposes; the sway of national governments ends at their borders.

In fact, nations are no longer fully sovereign even within their own borders. As a result of international trade agreements such as the World Trade Organization "WTO" Treaty, to which the US is a signatory, national laws can now be overturned almost at will by multinational firms if they can be shown, to the satisfaction of a WTO (or NAFTA the North American Free Trade Agreement) selected arbitration panel, to be an "unfair restraint on trade." Usually it is the provision that is "unfair" to the US that gets thrown out; it hardly matters what would be fair to, say, Ghana.

Across the world, national governments are less and less in control of their own policy-making decisions. Multinationals, many far richer than most nations, are increasingly in the driver's seat. More and more governments are becoming tools of this multinational juggernaut. These giant firms use their control of markets as well as their control of the regulatory functions of national governments to ensure that no competitors can emerge to challenge their economic fiefdoms.

At the level of the multinational firm, business is oligopolistic: for each market sector, only a few big firms dominate; Coke and Pepsi are examples. More to the point: a handful of companies such as Exxon-Mobil, Chevron Texaco, and Royal Dutch/Shell control the bulk of production in the oil and energy field. Adam Smithian "Economy 101" notions of free enterprise capitalism have no explanatory utility here.

In the US, the capture and effective subordination of the national government was appallingly easy. Money has always spoken in this country. But until recently, the political buying power of big business was more or less balanced by that of big labor. However, with the rise of multinational firms and the consequent deindustrialization of America, as decent-paying manufacturing jobs were exported to low-wage nations, this check is ever less effective. The industry-centered labor union movement has been gutted and national policy is ever more written by and for multinational firms.

Statistics show that in Congressional races, the candidate that spends the most money wins about 96% of the time. It's a case of "who has the gold makes the rules." The Supreme Court ruled in *Buckley v. Vallejo* (1976) that corporations have a right to inject essentially unlimited outside money into political campaigns (called "soft money"), as free speech rights that are guaranteed by the First Amendment.[200] Since then, the political system has become almost wholly beholden to the highest bidder, and no one has anywhere near as much money as the giant multi-nationals.

Although the Bipartisan Campaign Finance Reform Act of 2002 was upheld by a 5-4 majority of the US Supreme Court which ruled on December 10, 2003, nothing has fundamentally changed with respect to the relationship between money and politics in the US. Soft money was formally banned by the Act. However, a way was quickly found to circumvent the intent of the Act in banning soft money by means of what are called 527 groups. These groups get their designation from the section of the US Internal Revenue Code, section 527, which authorizes them. These groups can accept unlimited donations for political purposes so long as they do not specifically advocate voting for a particular candidate for elective office. The 2004 presidential election saw one such group, "Swift Boat Veterans for Truth," inflict major damage upon the candidacy of Sen. John Kerry without violating the letter of the law of the I.R.S. statute. So, nothing has really changed as a result of the Act's passage.

Government in the US was designed to be a *res publica*, a "thing of the people." However successfully we may have been advancing along the road towards that goal, it has become a republic "of, for, and by the corporation."

---

200. *Newshour With Jim Lehrer* Transcript, Oct. 5, 1999, A Question of Speech, http://www.pbs.org/newshour/bb/politics/financereform_10-5.html

Jefferson, arguably the most intellectual president, foresaw this outcome. Jefferson articulated his misgivings about the already growing power of corporations at the beginning of the 19th century:

> I hope we shall crush in its birth the aristocracy of our moneyed corporations which dare already to challenge our government in a trial of strength, and bid defiance to the laws of our country.[201]

So did President Abraham Lincoln. In a letter to Col. William F. Elkins, dated November 24, 1864, Lincoln presciently stated:

> We may congratulate ourselves that this cruel war is nearing its end. It has cost a vast amount of treasure and blood. . . . It has indeed been a trying hour for the Republic; but I see in the near future a crisis approaching that unnerves me and causes me to tremble for the safety of my country. As a result of the war, corporations have been enthroned and an era of corruption in high places will follow, and the money power of the country will endeavor to prolong its reign by working upon the prejudices of the people until all wealth is aggregated in a few hands and the Republic is destroyed. I feel at this moment more anxiety for the safety of my country than ever before, even in the midst of war. God grant that my suspicions may prove groundless.[202]

This wholesale capture of the US government by multinational corporations — and of governments across the planet — that was in progress in Lincoln's time is now very firmly entrenched. Corporate personhood is well entrenched in many nations besides the US. For example, in Canada:

> In Canada, the 1982 Canadian Charter of Rights and Freedoms offered no explicit personhood rights to corporations, but that didn't stop Canadian corporations from quickly mobilizing to make sure — via a number of successful court cases — that they were indeed defined as persons with some of the same Constitutional protections as actual persons.[203]

Corporate control of the mass media means we are constantly deluged with lies and omissions that further the corporations' self-interest and of course, the corporate owners' interests. Actual news is subordinated to the bottom line. Citizens have little ability to think critically. They accept at face value the misinformation fired at them daily.

Over a hundred years of increasingly effective marketing techniques, enhanced by an equal period of research by governments into what is colloquially called "brainwashing," the world's controlling elite have produced an

201. *Wikipedia, The Free Online Encyclopedia*: Corporate Personhood, Ref. 3rd paragraph, http://en.wikipedia.org/wiki/Corporate_personhood

202. Shaw, Archer H., *The Lincoln Encyclopedia*, Macmillian Inc, New York, NY, 1950, pp. 40, See also: http://www.ratical.org/corporations/Lincoln.html

203. Speech by Paul Cienfuegos, Director of Democracy Unlimited – presented at a major public forum on the USDA Organics Proposal: "Organic Agriculture in Jeopardy" at the Vancouver Planetarium in Vancouver B.C., Canada, May 18, 1998.

expertise that is now routinely used to "sell" to the public whatever policies they wish, via TV and the other corporatized media. Reality itself is now a "product" which is created and then mass marketed by these amoral elites.

This may sound hyperbolic to some readers. However, it is true. Dr. Norman Livergood was the Chair of the US Army War College's Artificial Intelligence Department between 1993 and 1995. Concerning his tenure there, Livergood states:

> While serving as Head of the Artificial Intelligence Department at the US Army War College for several years, I conducted studies on profiling, psychological programming, and brainwashing. I explored and developed personality simulation systems, an advanced technology used in military war games, FBI profiling, political campaigning, and advertising. Part of my discovery was that:
>
> • Unenlightened human minds are combinations of infantile beliefs and emotional patterns.
> • These patterns can be simulated in profiling systems.
> • These profiling systems can then be used to program and control people.[204]

> ....The way in which the Bush junta is conducting itself is an interesting brainwashing technique itself: Bush, Cheney, Ashcroft, Rumsfeld, and the others continually commit OUTRAGES but don't excuse them, explain them, or invite reflections on these affronts to morality and sanity. In fact, when some timid media voice criticizes the Bush junta, the person is demonized as questioning behavior which is beyond reproach. Americans are being brainwashed to ask only the questions the Bushites allow and they are programmed to see everything the Bush junta does as unquestionably correct.[205]

According to *Business Week*, the sophistication of computer programs designed by both government and private researchers to model and potentially to control humans is increasing exponentially:

> How do you convert written words into math? Goldman says it takes a combination of algebra and geometry. Imagine an object floating in space that has an edge for every known scrap of information. It's called a polytope and it has near-infinite dimensions, almost impossible to conjure up in our earthbound minds. It contains every topic written about in the press. And every article that Inform processes becomes a single line within it. Each line has a series of relationships. A single article on Bordeaux wine, for example, turns up in the polytope near France, agriculture, wine, even alcoholism. In each case, Inform's algorithm calculates the relevance of one article to the next by measuring the angle between the two lines. By the time you're reading these words, this very article will exist as a line in Goldman's polytope. And that raises a fundamental question: If long articles full of twists and turns can be reduced to a mathematical essence, what's next? Our businesses — and, yes, ourselves.[206]

---

204. Livergood, Norman D, *America Awake: We Must Take Back Our Country*, Dandelion Books LLC, Tempe, AZ, 2003, pps. 238-9.
205. Ibid., Livergood, Norman D., pp 238, # 204.

ECONOMIC COLLAPSE

A video entitled *The Corporation* compared the behavioral characteristics of multi-national corporations with those of sociopathic humans as defined by the *Diagnostic and Statistical Manual of Mental Disorders IV* (DSM) — which is the "Bible" for the psychologists and psychiatrists when diagnosing mental disorders. Curiously enough, corporate entities manifested every symptom of human sociopathology. The documentary's website summarizes this finding, stating that:

> To more precisely assess the "personality" of the corporate "person," a checklist is employed, using actual diagnostic criteria of the World Health Organization and the DSM-IV, the standard diagnostic tool of psychiatrists and psychologists. The operational principles of the corporation give it a highly anti-social "personality": It is self-interested, inherently amoral, callous and deceitful; it breaches social and legal standards to get its way; it does not suffer from guilt, yet it can mimic the human qualities of empathy, caring and altruism. Four case studies, drawn from a universe of corporate activity, clearly demonstrate harm to workers, human health, animals and the biosphere. Concluding this point-by-point analysis, a disturbing diagnosis is delivered: the institutional embodiment of laissez-faire capitalism fully meets the diagnostic criteria of a "psychopath."[207]

These entities are the foundation of the global and US national economies. They control political systems. Economic growth — meaning expansion — is their goal and this drives the political and economic system called globalization.

Consider the remarks of George W. Bush at a March 29, 2001 press conference, in which he renounced his campaign pledge to sign the Kyoto global warming accord: "I will explain as clearly as I can today and every other chance I get, that we will not do anything that harms our economy.... That's my priority. I'm worried about the economy."[208] Clearly, Mr. Bush was telling us that the bottom line of the corporations who had placed him in the office of President of the United States took precedence over all else. In doing so, he clearly reminded the people that their continued economic well being was contingent upon cooperation with their nation's hijackers.

Today's global economy grew out of the late medieval economy of Europe. The rapid expansion of European nations across the entire planet began with Columbus's 1492 discovery of the Americas led quickly to its becoming global in extent. With the colonization of various parts of the world by various nations in

---

206. *Business Week* Online, Jan 23, 2006, Math Will Rock Your World, http://www.businessweek.com/magazine/content/06_04/b3968001.htm?campaign_id=nws_insdr_jan13&link_position=link1

207. The Corporation: A Film by Mark Achbar, Jennifer Abbot, and Joel Bakin, http://www.thecorporation.com/

208. American Institute of Physics, FYI: The AIP Bulletin of Science Policy News, FYI Number 39, April 5, 2001, Global Climate Change: Bush Remarks and New Report, http://www.aip.org/fyi/2001/039.html

Europe, what was hitherto a mostly self-contained regional European economy suddenly and explosively expanded to become a world economy.

Private corporations were chartered by national governments to exploit the riches of the nations which fell before their world-wide expansion. The British East India Company, chartered in 1600, along with the Dutch East India Company, which was chartered in 1602, were the world's first multinational corporations.[209] [210] Their reason for existence was to exploit the human and material resources of the planet in order to enrich their stockholders as much as possible, as quickly as possible. The mandate of global corporations remains unchanged today. The only real change in the past four hundred years is that they now control human governments rather that being controlled by those governments.

Energy equals the ability to do work. Work generates wealth. Wealth is denominated by money. Economic expansion today is paid for by loans which represent a bet that new wealth, meaning new energy, will be available in the future to pay back the loan with interest.

> The map is not the territory. To any thinking person this statement is axiomatic. But, an important corollary is more difficult to discern, namely: *Money is nothing more than the right to command energy to do what you want it to do.* To understand how the two statements above go together you need first to understand that money and credit make up what's called the symbolic economy. The symbolic economy merely *represents* what is happening in the real economy of goods and services including energy goods. Second, it is useful to know that only a fraction of one percent of all energy which goes into the products and services of an industrial economy comes from physical human labor. All the rest comes from a mix of fossil fuels (86%), nuclear power (7%), hydro power (6%) and alternative sources (1%). Since nothing gets mined, grown, harvested, processed, manufactured or delivered without energy, it follows that energy is the true currency of modern civilization. And, without energy sources that go beyond human and animal labor, we would revert to a pre-industrial lifestyle.[211]

This really restates to an assumption that nature is an essentially infinite repository of energy and therefore of wealth, which can be exploited without limit and without fear. If this assumption fails, economic collapse must occur, and occur rapidly.

> Ancient discovery of fire and the possibility of burning wood made available, for the first time, fairly large amount of energy for mankind. Later (4000 and 3500 years B.C.) after the first sailing ships and windmills were developed and the use of hydropower began via water mills or irrigation systems, cultural development

209. *Encyclopædia Britannica*, 2006., Dutch East India Company, Encyclopædia Britannica Premium Service, http://www.britannica.com/eb/article-9031608.

210. *Encyclopædia Britannica*, 2006, East India Company, Encyclopædia Britannica Premium Service, http://www.britannica.com/eb/article-9031775.

211. *Energy Bulletin*, Dec. 5, 2005, Money and energy: The map is not the territory, http://energybulletin.net/11460.html

began to accelerate. For several thousands years human energy demands were covered only by renewable energy sources — sun, biomass, hydro and wind power. It was only until the start of industrial revolution and the ability to transform heat into motion, when energy consumption and industrial development accelerated rapidly. The industrial revolution was a revolution of energy technology based on fossil fuels. This occurred in stages, from the exploitation of coal deposits to oil and natural gas fields on a global scale. It has been only half a century since nuclear power began being used as an energy source.[212]

Oil production per capita peaked in 1979. Per capita oil production has been declining ever since, because human population has increased at a much greater rate than oil production has increased. Writing in 2001, researcher Richard C. Duncan, a PhD electrical engineer and the Director of the Institute on Energy and Man, summarized this relationship as follows:

> Although world oil production from 1979 to 1999 increased at an average rate of 0.75%/year, world population (Pop) grew even faster. Thus world oil production per capita declined at an average rate of 1.20 %/year during the 20 years from 1979 to 1999.[213]

The situation is slightly more complicated when all sources of energy — not just oil — are taken into account. Writing in 2005, Duncan found that when all energy sources including natural gas, hydroelectric, and nuclear are taken into account, net energy per capita growth was zero during the 1979–2003 period for which data were available.[214]

Given that energy is wealth, it must follow that real wealth per capita must have stagnated or decreased globally during this period as well. This is nowhere more true than when sorting through the forests of often politically skewed statistics dealing with income. Researchers from the Hudson Institute went right to the heart of the matter: "No matter how we fiddle with the numbers, real income growth has still slowed dramatically, to virtually zero in recent years.[215]

Income inequality has increased dramatically:

> In 2004, after three years of economic recovery, the US Census reports that poverty continues to grow, while the real median income for fulltime workers has declined.

212. Energy Saving Now, Why do we need the change in energy use and production?, http://energy.saving.nu/energytoday/consumption.shtml
213. Duncan, Richard, C., Institute on Energy and Man, World Energy Production, Population Growth, and the Road to the Olduvai Gorge, As published in *Population and Environment*, May-June, 2001, v. 22, n. 5, http://www.hubbertpeak.com/duncan/road2olduvai.pdf
214. *The Social Contract*, Vol. 16, No. 2, Winter 2005-2006, The Olduvai Theory: Energy, Population, and Industrial Civilization, Richard C. Duncan, pp. 6.
215. *Facts & Opinions*, May 1997, Just the Facts — About Income Growth, Volume 3, Number 2, http://66.102.7.104/search?q=cache:XtnHuz76300J:www.limitedgovernent.org/publications/pubs/FandO/fomay97.pdf+income+stagnation&hl=en

Since 2001, when the economy hit bottom, the ranks of our nation's poor have grown by 4 million, and the number of people without health insurance has swelled by 4.6 million to over 45 million. Income inequality is now near all-time highs, with over 50 percent of 2004 income going to the top fifth of households, and the biggest gains going to the top 5 percent and 1 percent of households. The average CEO now takes home a paycheck 431 times that of their average worker. At the pinnacle of US wealth, 2004 saw a dramatic increase in the number of billionaires. According to Forbes Magazine, there are now 374 US billionaires. The growth in billionaires took a dramatic leap since the early 1980s, when the average net worth of the individuals on the Forbes 400 list was $400 million. Today, the average net worth is $2.8 billion. Wal-Mart's Walton family now has 771,287 times more [wealth] than the median US household.[216]

Statistics can obfuscate but can't completely conceal the simple truth of the 1979 to present period. Since 1979, real incomes for Americans have stagnated and have even declined, particularly for the poorest. American households now consist of more workers, working longer hours for lower wages. At the same time, ever more rapid concentrations of wealth in the hands of a miniscule percentage of the population has intensified. The corporate-connected rich are rapidly getting richer; the poor are getting poorer, while the former middle classes are ever more squeezed.

Another way to get a handle on the increasing income stratification in the US is to look at median domestic income. This is the income received by the 50th percentile of income earners. This statistic is important, as the mean income does not tell us how equally income is distributed. For example, if there were ten income earners and one earned $1,000,000.00 while the rest earned only $1,000.00 per year, the mean or average income for this group would be $100,900.00, while the median income would be 1,000.00. Clearly since nine out of ten people received incomes of only $1,000.00 in my example, median income best captures income inequality.

As Mark Twain once observed: "There are three kinds of lies: lies, damn lies, and statistics."[217] In this vein, Former United States Treasury Secretary Robert B. Reich states:

> Listen to most economic commentators and you'd think the biggest news of 2005 was that the American economy continued to grow at a healthy clip — notwithstanding hurricanes, oil shocks, trade imbalances and a bloated federal budget deficit. Well, that's true. The power and resilience of this economy are remarkable. But there's another story about the American economy that's equally remarkable, although more sobering. Although the data aren't all in, it seems almost certain that in 2005, median incomes continued to drop. It's been that way for four years now, since the end of the last recession. The economy keeps growing, but median

---

216. Alonovo.com, Oct. 20, 2005, Inequality in America: Version 2.0, http://www.alonovo.com/community/node/42

217. Click Z Network: Solutions for Marketers, Aug. 13, 2002, Lies, Damn Lies, and Statistics, http://www.clickz.com/experts/design/freefee/article.php/1445011

incomes keep declining. Take inflation into account and you find that half of all American workers are earning less now than they did in 2001. Rarely before in history has there been such a long period of growth in the gross domestic product without most Americans sharing in that growth. Forget "trickle-down" economics. Even if you believe the Bush tax cuts of 2001 and 2003 helped the economy grow — and if you do, you probably believe in Santa Claus — nothing is trickling down, not even to the middle. Most is going to the top fifth. And most of that is going to the top 5 percent.[218]

Reich's observation also neatly illustrates the propagandistic nature of the corporate-consolidated American mass media, which uses the "damn lies" to propagate the notion that all the proverbial boats are rising (even though many of them are sinking). For example, a recent study says that:

New government data indicate that the concentration of corporate wealth among the highest-income Americans grew significantly in 2003, as a trend that began in 1991 accelerated in the first year that President Bush and Congress cut taxes on capital. In 2003 the top 1 percent of households owned 57.5 percent of corporate wealth, up from 53.4 percent the year before, according to a Congressional Budget Office analysis of the latest income tax data. The top group's share of corporate wealth has grown by half since 1991, when it was 38.7 percent. In 2003, incomes in the top 1 percent of households ranged from $237,000 to several billion dollars. For every group below the top 1 percent, shares of corporate wealth have declined since 1991. These declines ranged from 12.7 percent for those on the 96th to 99th rungs on the income ladder to 57 percent for the poorest fifth of Americans, who made less than $16,300 and together owned 0.6 percent of corporate wealth in 2003, down from 1.4 percent in 1991.[219]

Despite all of the media-disseminated happy talk about the economy, the very rich are becoming very much richer at an increasing rate. Wealth is being concentrated into fewer and fewer hands.

Since additional hydrocarbon energy is unavailable to expand the economy, jobs are outsourced to lower wage areas in order to keep profits up. The basic reality underlying all of these economic changes is declining energy (which equals wealth) per capita. This is occurring at the same time as the consolidation of control over the political structures of the US and other nations by multinational corporations in general, and specifically by the global energy companies.

The new ruling class can and does utilize their ownership of the media to control the beliefs and actions of the population. In general, people now have their reality manufactured for them by the corporatists.

---

218. *TomPaine.Com*, Jan. 3, 2006, Our Worrisome MDP, http://www.tompaine.com/articles/20060103/our_worrisome_mdp.php

219. *New York Times*, Jan. 29, 2006, Corporate Wealth Share Rises for Top Income Americans, http: //www.nytimes.com/2006/01/29/national/29rich.html?_r=1&adxnnl=1&oref=slogin&adxnnlx=1138729465-4Z2MDlihO3/BhE4nnRtE9Q

In his 2005 article "The Olduvai Theory, Energy, Population, and Industrial Civilization," Richard C. Duncan, presents this projected decline:

> The Olduvai Theory states that the life expectancy of industrial civilization is approximately 100 years: circa 1930-2030. Ackerman's ("White's") Law defines it: $e =$ *Energy/Population*. Four postulates follow:
>
> 1. The exponential growth of world energy production ended in 1970.
>
> 2. Average e will show no growth from 1979 to circa 2008.
>
> 3. The rate of change of e will go steeply negative circa 2008.
>
> 4. World population will decline proximate with e.

Universally accepted data confirm Postulate 1. The data are in accord with Postulate 2 up until 2003, the most recent year for which comprehensive data are available. Postulate 3 remains to be seen, but the evidence is pointing that way.

Duncan does not take climate effects of hydrocarbon usage or political failure to respond to peak oil and climate change into account. These factors suggest that the rate of population decline will actually be *steeper* than that projected in Postulate 4 of his Olduvai theory. In other words, world population could decline even faster than does e.[220]

Joseph Tainter, in his masterful work *The Collapse of Complex Societies*,[221] describes how complex societies — civilizations — when they collapse, are reduced in geographic extent, often breaking up into multiple smaller, less complexly organized successors. These entities, being simpler, use substantially less energy. In effect, a sudden decrease in available energy due to war, famine, drought, flood, or other internal or external shocks to the civilization's political economy causes it to contract and break up into simpler, less energy-intensive modes of organization; this allows for life to continue, for any who survive, on a reduced energy budget.

Collapses of civilizations in the past have involved localized failures of primarily agrarian-based civilizations. They have generally resulted in a mortality rate of about ninety percent.

Corporate artificial "persons" exploit Americans' nearly unthinking loyalty to their institutions, particularly the nation-state, and also, for many, to religious institutions. Throw in corporate control of the media and it's easy to see that the majority of the public do not even know what are their own interests. An informative book about corporate personhood and its lethal implications is Thom Hartman's *Unequal Protection: The Rise of Corporate Dominance and the Theft of Human Rights*.[222]

---

220. Ibid., Duncan, Richard C., # 213.

221. Tainter, Joseph, *The Collapse of Complex Societies*, Cambridge University Press, Cambridge, UK, 1988.

222. Ibid., Hartman, Thom, # 194.

CHAPTER 8. PROPAGANDA, BETRAYAL, AND WAR

There is danger from all men. The only maxim of a free government ought to be to trust no man living with power to endanger the public liberty. — John Adams [223]

WARS AND RUMORS OF WARS

Multinational corporations have an interest in controlling national populations in order to motivate them to willingly sacrifice, suffer, and even die for the profit of the multinationals. Advancing corporate agendas by obtaining public-supported political consent for them has became a core function of the media, and of the corporate-controlled political system.

Since national and global political systems have been centered upon these multinationals, the global properties of these nested systems quickly manifested emergent properties which facilitated the agendas of all those whose goals that happened to further enhance corporate power and rule. This was not the work of a secret nefarious cabal of conspirators. Rather it was simply the organizing power of the transformed systems. Emergence causes this type of spontaneous ordering, almost as a magnet causes iron filings to line up in a precise pattern — without any premeditation. As this was occurring, corporate propaganda memes, justifying all their actions, have created a distorted sense of reality in the minds of most of the population. Creating "reality" came to be an essential marketing and societal control tool.

This has made it ever more difficult to interpret the news as it is fed to us. The so called "War on Terror" is actually a war for control of oil and other resources and their geo-strategic transit routes, as discussed above. In fur-

---

223. Brainy Quote: John Adams Quotes, http://www.brainyquote.com/quotes/authors/j/john_adams.html

therance of these goals, it is also a war against citizen control over the policies of the nation-state. This is due to a simple reality: The policies that benefit ordinary citizens are different from those that benefit multi-national corporations and their majority shareholders. The aim is to further consolidate the power of the corporate owners. Of course, such policies are disempowering and impoverishing to the rest.

People are not willing to die for the interests of a corporation. However, they are willing do so for their nation. Therefore, nation-states are used to promote and legitimate the policies that work for corporations. The trick is to substitute the national interest for the corporate interest without the sacrificial victim becoming aware of this. George W. Bush actually blurted out on one occasion:

> See, in my line of work, you got to keep repeating things over and over again for the truth to sink in, to kind of catapult the propaganda.[224][225]

The wealthiest and most powerful corporations derive their wealth and power from extracting, refining and distributing hydrocarbon derived products. Accordingly, any threats to business as usual, such as declining reserves of hydrocarbons, are direct threats to these elites continued wealth and power.

Consequently, we were subjected to a sophisticated marketing campaign to promote a war for control over the world's remaining energy reserves by the big energy multi-nationals who controlled the Bush Administration.

CORPORATISM, FASCISM, AND STOLEN ELECTIONS: PIECES OF A PUZZLE

Americans were told that they had been subjected to a violent, unprovoked attack, for which there had been no warning, by Middle-Eastern terrorists who "hate our freedoms." We all saw the flaming World Trade Center towers come crashing down, and we saw the Pentagon burning on September 11, 2001. Americans instinctively rallied around the corporatists' point men, President George W. Bush and Vice-President Richard Cheney.

I use the term "corporatist" throughout this work to refer specifically to the congruence of state power with corporate power. In this system of political economy the interests of state and corporation become effectively unified. The state becomes an instrument by which corporate goals are implemented. This occurs, in my definition of corporatism, when corporate power captures and

---

224. The White House, Office of the Press Secretary, May 24, 2005, President Participates in Social Security Conversation in New York, http://www.whitehouse.gov/news/releases/2005/05/20050524-3.html

225. Political Humor: Top 10 Bushisms of 2005, http://politicalhumor.about.com/od/bushquotes/a/topbushisms2005.htm

subsumes the state. The state then becomes a tool of an elite class of corporate directors.

Additionally, corporatism is a necessary, though not sufficient, condition for fascism as I define it in this work. Fascism requires a corporatist state. Fascism is characterized by belligerent nationalism; concentration of power into the hands of a leader who is effectively above the law and meaningful public criticism; *de jure* or *de facto* control of the media; and either no elections, or rigged elections.

A continuum exists where corporatism shades imperceptibly into fascism. Since 9/11 the US has been generating unending barrages of belligerent nationalism. Bush's message is very simple: "You are either with us or against us."[226] [227] In an address to the US Congress Bush slightly restated this maxim: "Either you are with us, or you are with the terrorists." In other words, if you exercise your rights and fulfill your duty as a responsible citizen by examining and questioning the policies and actions of your country, you are disloyal. You are practically a terrorist yourself.

This same President has also unilaterally given himself the right to deal in any way he chooses with people caught up in the dragnet for terrorists, questioners, and scapegoats — and he alone gets to decide who fits this category. Amazingly, rather than protesting at this gross violation of the US Constitution and its Bill of Rights, the mainstream American media has been cheerleading it! Such behavior certainly moves us far along the spectrum from corporatism to full-blown fascism.

The historical record since 9/11 strongly indicates an out-of-control, above-the-law, and ever more powerful executive branch within the US. The craven submission of the mainstream media during this period is also only too evident.

What about elections? Bush came to power in 2000 as a result of a highly partisan 5-4 decision of the US Supreme Court. That was an obviously political decision, and the Supreme Court's very intervention flagrantly violated Article II of the US Constitution, which plainly states that:

> Each state shall appoint, in such manner as the Legislature thereof may direct, a number of electors, equal to the whole number of senators and representatives to which the State may be entitled in the Congress: but no Senator or Representative, or person holding an office of trust or profit under the United States, shall be appointed an elector.[228]

226. The White House, Address to a Joint Session of Congress and the American People, Sept. 2001, http://www.whitehouse.gov/news/releases/2001/09/20010920-8.html

227. *CNN.com*, Nov. 6, 2001, War on Terror, 'You are either with us or against us', http://archives.cnn.com/2001/US/11/06/gen.attack.on.terror/

228. United States Constitution, Cornel Law School, http://www.law.cornell.edu/constitution/constitution.overview.html

It is for the states themselves (such as Florida) to appoint electors. The national government has no role in this matter whatsoever. The Supreme Court violated Article II of the US Constitution when it intervened in Florida's Presidential elector selection process.

As to the 2004 US Presidential election, most people probably do not realize that exit polls showed Bush's opponent, John Kerry, to have been the clear winner of the election with a margin of over 3 million popular votes and more than 100 Electoral College votes. Research by professional statisticians such as Dr. Steven Freeman and many others demonstrate that there can be little doubt that these polls — which are interviews of people who have just voted — were accurate. What was inaccurate was the publicly certified vote count that gave Bush the Presidency.

These statisticians conclude that:

> The exit pollster of record for the 2004 election was the Edison/Mitofsky consortium. Their national poll results projected a Kerry victory by 3.0%, whereas the official count had Bush winning by 2.5%. Several methods have been used to estimate the probability that the national exit poll results would be as different as they were from the national popular vote by random chance. These estimates range from 1 in 16.5 million to 1 in 1,240. No matter how one calculates it, the discrepancy cannot be attributed to chance.[229]

In their peer-reviewed paper, they conclusively eliminate any possibility that the discrepancy can be attributed to sampling error, chance, or inaccurate exit polls.[230]

The most reliable statistical assumptions concerning the discrepancies for the 2004 US Presidential election are in the range of anywhere between hundreds of thousands to one, to millions to one, in favor of a fraudulent outcome. The lowest possible estimates for the 2004 elections results discrepancy versus the exit polling are well over an order of magnitude greater than California's legal standard for judicially establishing paternity or guilt of a crime.

Environmental attorney Robert F. Kennedy Jr. wrote in the June 1, 2006 edition of *Rolling Stone* an article entitled "Was the 2004 Election Stolen?" The magazine had undertaken an extensive four-month-long investigation with Kennedy, focused upon Ohio, the state in which the apparent fraud was most egregious. They found massive, well-orchestrated vote fraud of almost every type conceivable. The magnitude of the fraud was far more than enough to have stolen the state of Ohio for Bush and away from Kerry, who beyond any reasonable

---

229. US Counts Votes National Election Data Archive Project, Analysis of the 2004 Presidential Election Exit Poll Discrepancies, March 31, 2005, Updated April 12, 2005, http://www.uscountvotes.org/ucvAnalysis/US/Exit_Polls_2004_Edison-Mitofsky.pdf
230. Peer-reviewed research by these statisticians, *which has never been refuted*, can be found at the US Counts Votes.org website at: http://www.uscountvotes.org/.

doubt won the state — and the Presidency — by a substantial margin in 2004. Kennedy says:

> After carefully examining the evidence, I've become convinced that the president's party mounted a massive, coordinated campaign to subvert the will of the people in 2004. Across the country, Republican election officials and party stalwarts employed a wide range of illegal and unethical tactics to fix the election. A review of the available data reveals that in Ohio alone, at least 357,000 voters, the overwhelming majority of them Democratic, were prevented from casting ballots or did not have their votes counted in 2004 — more than enough to shift the results of an election decided by 118,601 votes. In what may be the single most astounding fact from the election, *one in every four* Ohio citizens who registered to vote in 2004 showed up at the polls only to discover that they were not listed on the rolls, thanks to GOP efforts to stem the unprecedented flood of Democrats eager to cast ballots. And that doesn't even take into account the troubling evidence of outright fraud, which indicates that upwards of 80,000 votes for Kerry were counted instead for Bush. That alone is a swing of more than 160,000 votes — enough to have put John Kerry in the White House.[231]

Kerry won Ohio. Additionally, massive, systematic, and organized fraud occurred nationwide. Kerry, not Bush, won the election in 2004.

This explosive story was never reported upon by the mainstream media, except briefly to dismiss it as being beyond the pale of permissibility. This included the so-called "liberal media." *The Washington Post*, for example, was quick to dismiss any and all allegations of fraud as "conspiracy theories," while *The New York Times* swiftly declared that "... there is no evidence of vote theft or errors on a large scale."[232] And with the lonely exception of Keith Olberman who continued for several weeks to report on the story on his weeknight show *Countdown*, on MSNBC, that was that.

This is a real world case of "the dog that didn't bark." In the Sherlock Holmes short story *Silver Blaze*, one seemingly small detail solves the crime: a dog that fails to bark. Why? Because this dog recognizes the murderer — it's his owner.

Similarly, the corporate-owned press fails to "bark"; that is, to alert us to misuses and abuses of power on the part of corporate interests, whether in elections or in drumming up support for a war.

Furthermore, whenever the policies implemented by one set of corporatists become deeply unpopular, they can be "voted" out of office by the people and replaced with another set of names and faces — who end up representing the same interests as the group that was just dismissed.

Senator John Kerry himself did not protest the charade. By all accounts John Kerry was a brave soldier in combat in Vietnam, and afterwards in political

---

231. *Rolling Stone*, June 1, 2006, Was the 2004 Election Stolen?, http://www.rollingstone.com/news/story/10432334/was_the_2004_election_stolen/1
232. Ibid., *Rolling Stone*, # 231

combat in protest of that war. However, discretion is the better part of valor and in this case he seems to have found the risks too high and the chance of making a difference too negligible; he preferred to live to fight another day.

According to Professor Mark Crispin Miller[233], author of *Fooled Again, How the Right Stole the 2004 Election & Why They'll Steal The Next One Too (Unless We Stop Them)*, Senator Kerry admitted to him in a private conversation that he knew that the 2004 Presidential election had been stolen from him. A spokesperson for Kerry later denied that he had said this however. Recounting these events, Miller stated:

> Kerry "told me he now thinks the election was stolen. He says he doesn't believe he is the person that can be out in front because of the sour grapes question. But he said he believes it was stolen. He says he argues with his democratic colleagues on the hill. He said he had a fight with Christopher Dodd because he said there's questions about the voting machines and Dodd was angry." Miller was shocked to hear of Kerry's denial.[234]

Some Kerry supporters wanted to believe he was the new "Kennedy." Is it possible he, and Gore too, wanted to avoid the Kennedy fate? Or, as Gore and others have suggested, are they concerned that if they speak out too strongly and the truth about the stolen elections of 2000 and 2004 become widely known, the American political system will have be irrevocably de-legitimated? Those who have benefited from these elections will never concede voluntarily — and they are backed fully by all of the power of the multi-national corporations, and by the media's deceptive powers.

Curiously, at almost the very moment that the American presidential election was being stolen — again — an eerily similar theft was attempted in the Ukrainian presidential election. The primary forensic evidence for the Ukrainian election theft was, again, the discrepancy between exit poll results and official election results. The 2006 election in Mexico is also being rejected by the public as having been totally compromised.

The falsification of election results in the US is facilitated by the fact that the overwhelming majority electronic voting machines are manufactured by a handful of Republican-connected corporations. Researchers have demonstrated over and over again that these devices can be hacked or rigged to deliver any desired outcome for an election without leaving a trace of the crime. Because there is no paper trail from the devices, a meaningful recount is impossible. Further, US courts have ruled that the software used by the machines is proprietary and that therefore elections departments in the various states cannot have

---

233. See his website at: http://www.markcrispinmiller.blogspot.com/ for additional details.
234. *The Raw Story*, Nov. 4, 2005, Senator Kerry Rebuffs Claim He Said Election Was Stolen, http://rawstory.com/news/2005   Senator_Kerry_rebuffs_claim_he_said _1104. html

access to it. Elections officials are forced to take the manufacturer's word that the vote will be fair and accurate.[235]

The Government Accountability Office (GAO) investigated complaints about electronic voting in a report to the US Congress released in September 2005. In that report the GAO stated that:

> While electronic voting systems hold promise for a more accurate and efficient election process, numerous entities have raised concerns about their security and reliability, citing instances of weak security controls, system design flaws, inadequate system version control, inadequate security testing, incorrect system configuration, poor security management, and vague or incomplete voting system standards, among other issues. For example, studies found (1) some electronic voting systems did not encrypt cast ballots or system audit logs, and it was possible to alter both without being detected; (2) it was possible to alter the files that define how a ballot looks and works so that the votes for one candidate could be recorded for a different candidate; and (3) vendors installed uncertified versions of voting system software at the local level. It is important to note that many of the reported concerns were drawn from specific system makes and models or from a specific jurisdiction's election, and that there is a lack of consensus among election officials and other experts on the pervasiveness of the concerns. Nevertheless, some of these concerns were reported to have caused local problems in federal elections — resulting in the loss or miscount of votes — and therefore merit attention.[236]

So not only does contemporary America feature belligerent nationalism trumpeted by a corporate-selected leader and a tame and generally obedient press, it also features corporate-controlled elections using devices which are run by secretive software, are easily hacked or rigged, and whose results cannot be audited or recounted.

IN WHOSE INTEREST?

On September 11, 2001, the separation of corporate interest and national interest was breached.

Amidst all of the rallying to the leaders, the media and the shocked, mind-numbed mass public failed to ask why, with more than one hour's warning that commercial aircraft were being hijacked, the most heavily defended building on the planet, the Pentagon, had been left completely undefended. Or for that matter, if it was indeed a "screw-up" as the Bush Administration claimed, why

---

235. For further information, and research on electronic voting, readers may refer to Black Box Voting's website at: http://www.blackboxvoting.org/.

236. Report to Congressional Requestors, Sept. 2005, Federal Efforts to Improve Security and Reliability of Electronic Voting Systems Are Under Way, but Key Activities Need to be Completed, http://www.democrats.reform.house.gov/Documents/20051021122225-53143.pdf

no one was ever punished, reprimanded, fired, presumably not even scolded for this "oversight."

No credible explanation for the failure to provide air cover over the Pentagon has ever been proffered by the US government. The 9/11 Commission's account,[237] which concluded that many mistakes were made, ignored or disregarded mountains of evidence, as researcher David Ray Griffin has carefully documented.[238]

Decision makers in the Bush Administration, including Vice President Richard Cheney, who appears to have been in overall command of events on that tragic day, were signatories on the 1998 report by the Project for a New American Century (PNAC) titled, "Rebuilding America's Defenses."[239] In that report, PNAC outlined, in writing, and publicly, a blueprint for permanent military hegemony across the world which would facilitate its domination by US-based multi-national corporations in perpetuity.

Apparently realizing that such a sweeping change in policy would be difficult to achieve, the PNAC states on page 52 of this report that their agenda could not be implemented *absent some catastrophic catalyzing event — like a new Pearl Harbor.*

Signatories to the report include Vice President Richard Cheney; Defense Secretary Donald Rumsfeld; the then Deputy Defense Secretary and current World Bank head, Paul Wolfowitz; Ambassador (Proconsul) to Iraq Zalmay Khalizad (previously ambassador to Afghanistan); UN Ambassador John Bolton; and a host of lesser known neoconservatives in and out of government. The PNAC report *Rebuilding America's Defenses*, its contents, and the identities of its signatories are not disputed by anyone—including those now prominent Bush Administration figures who signed it in 1999. Prof. Peter Philips has observed:

> Among the PNAC founders were eight people affiliated with the number-one defense contractor Lockheed-Martin, and seven others associated with the number-three defense contractor Northrop Grumman. Of the twenty-five founders of PNAC twelve were later appointed to high level positions in the George W. Bush administration.[240]

---

237. The 9/11 Commission Report: Final Report of the National Commission on Terrorist Attacks Upon the United States, Authorized Edition, W.H. Norton and Company Inc., New York, NY, 2004.

238. Griffin, David Ray, *The New Pearl Harbor: Disturbing Questions About the Bush Administration and 9/11*, Olive Branch Press, New York, NY, 2006. See also this article: *Information Clearinghouse.com*, The 9/11 Commission Report: Omissions and Distortions, Prof. David Ray Griffin, http://www.informationclearinghouse.info/article8765.htm.

239. *Rebuilding America's Defenses: Strategy, Forces and Resources for a New Century*, A Report of the Project for the New American Century, Sept. 2000, http://www.newamericancentury.org/RebuildingAmericasDefenses.pdf

240. *Common Dreams News Center*, Feb. 9, 2006, Is US Military Domination of the World A Good Idea? http://www.commondreams.org/views06/0209-32.htm

This group articulated a need for a new Pearl Harbor type event in order to consolidate their agenda for corporatist dominion both within America and across the planet. This catalyzing event obligingly occurred within eight months of their taking control of the executive branch of the US government.

Like Pearl Harbor, there is considerable disagreement as to whether 9/11 was truly the result of "errors," was simply allowed to happen, or was in some way assisted by someone within the Bush administration. Either way, what has followed the 9/11 attacks has been a geo-strategically oriented war for control over vital energy resources and transport routes. It is irrefutable is that the Bush Administration exploited the 9/11 tragedy to its fullest extent in order to advance their pre-existing PNAC conceived agenda.

David Ray Griffin, who is mentioned above, along with Michael C. Ruppert, have written in great detail on these events in their books, *The 9/11 Commission Report: Omissions and Distortions* and *Crossing the Rubicon: the Decline of the American Empire at the End of the Age of Oil*.[241]

In the immediate aftermath of this "new Pearl Harbor," another agenda item was implemented — the further consolidation of control over the political system. Within days, a 342-page piece of legislation appeared. It was the USA PATRIOT Act, which effectively annuls the Fourth, Fifth, Sixth, and portions of the First Amendments to the US Constitution's Bill of Rights. Officially, Assistant Attorney General Viet D. Dinh and current Secretary of Homeland Security Michael Chertoff were depicted as its primary drafters. In actuality, given its staggering complexity and the short time in which it appeared in its full form, it must have been largely prepared well in advance of 9/11, and not in response to 9/11.

Senate Minority Leader Tom Daschle, working in coordination with Senate Judiciary Committee Chair Patrick Leahy, whose committee had to vote to approve the bill in order for it to move to consideration by the full Senate, tried to allow time for careful analysis of its contents. Perhaps coincidentally, in early October of 2001, while the PATRIOT Act remained stalled, anthrax-contaminated letters were mailed to both of them:

> The letters were addressed to two Democratic senators, Tom Daschle of South Dakota and Patrick Leahy of Vermont. More potent than the first anthrax letters, the material in the Senate letters was a highly refined dry powder consisting of approximately one gram of nearly pure spores. Some reports described the material in the Senate letters as "weaponized" or "weapons grade" anthrax. The Daschle letter was opened by an aide on October 15, and the government mail service was shut down. The unopened Leahy letter was discovered in an impounded mail bag on November 16. The Leahy letter had been misdirected to the State Department mail

241. Ruppert, Michael C., *Crossing the Rubicon: the Decline of the American Empire at the End of the Age of Oil*, New Society Publishers, Gabriola Island, B.C., Canada, 2004; Griffin, David Ray, *The 9/11 Commission Report: Omissions and Distortions*, Olive Branch Press, Northampton, MA, 2005.

annex in Sterling, Virginia, due to a misread Zip code; a postal worker there, David Hose, contracted inhalation anthrax.[242]

Daschle and Leahy withdrew their opposition and the USA PATRIOT Act was rushed through Congress. On October 25th, ten days after the Daschle anthrax-contaminated letter was opened, the full Senate passed the legislation with a vote of 98-1, with Wisconsin Senator Russell Feingold casting the lone dissenting vote. Bush signed it into law the following day.

Horrified and frightened by the events of 9/11 and the anthrax attacks, the public put up no substantial opposition to the Act. Legislation which abrogated the Constitutional freedoms was passed, largely unread, almost unanimously, by an intimidated Congress. The process was not entirely different from the manner in which the Reichstag passed the Enabling Act on March 23, 1933 that effectively handed Adolf Hitler all power in Germany.

At the same point in time in early 2006 when the renewal of the USA PATRIOT Act was being stalled by a group of relatively moderate Republican senators, the Russell Senate Office Building was evacuated due to what was described as being a nerve agent alarm being triggered. One of the leaders of these moderate Republican senators, Senator Chuck Hagel of Nebraska, was actually quarantined by officials wearing hazardous material suits during this incident.

Such alarms are almost unheard of. The last major biohazard scare in the Capitol was the anthrax attacks about four years ago,[243] coincident with the passage of the PATRIOT Act. Opposition to renewal of the Act collapsed the next day. On February 10 the Washington Post reported that:

> Efforts to extend the USA Patriot Act cleared a major hurdle yesterday when the White House and key senators agreed to revisions that are virtually certain to secure Senate passage and likely to win House approval, congressional leaders said.[244]

On February 17, all resistance to the "compromise" version of the Act — which contained all of the provisions demanded by the Bush Administration — collapsed as the Senate voted 97-3 to end further debate on the Act:

> The US Senate overwhelmingly voted to end a filibuster and moved closer to renewal of the USA Patriot Act. Senators voted 96-3 Thursday to stop debate regarding a compromise on the Patriot Act. All three of the senators who voted to keep debate going were Democrats and include Sen. Russ Feingold, D-Wis., who

242. *Encyclopædia Britannica*, 2006, United States, Encyclopædia Britannica Premium Service, http://www.britannica.com/eb/article-222383.
243. *ABC News*, Feb. 8, 2006, Security scare forces US Capitol evacuation, http://abcnews.go.com/US/wireStory?id=1596653&page=1ireStory?id=1596653&page=1
244. *Washington Post*, Feb. 10, 2006, page A1, Patriot Act Compromise Clears way for Senate Vote, http://www.washingtonpost.com/wp-dyn/content/article/2006/02/09/AR2006020902140.html?nav=rss_email/components

was the only senator to vote against the Patriot Act in 2001. He described recent changes in the bill as cosmetic.[245]

On March 9, 2006, the 16 expiring sections of the Patriot Act were signed into law by President Bush, and 14 of the 16 sections were made permanent by the renewal legislation. The Bush Administration had achieved all of their goals even as many Democratic legislators hailed the renewal as a compromise.

Furthermore, the apparent use of weapons grade anthrax containing a strain of that bacillus created at the US National Bioweapons Laboratory at Fort Detrick Maryland[246] [247] raises dire questions. The evidence is circumstantial and it cannot be ruled out that the perpetrator was an ideologically motivated "lone wolf." It is, however, certainly domestic — not foreign — terrorism.

The fundamental duty of government is to protect its citizens. Something like the Patriot Act or certain portions of it could serve that goal. There are instances in which different government agencies should be allowed to share information with one another. I am a former US Federal Agent, and in my capacity as an I.R.S. Revenue Officer there were cases in which I was unable to track down a tax delinquent. I was reasonably certain that the Social Security office on the floor below I.R.S. headquarters had up-to-date contact information about the person. However, I was forbidden by law from requesting this information.

This said, the bulk of the Act — including those 16 provisions which were renewed or made permanent in March 2006 — represented an excessive and unwarranted transfer of power from citizens to government, and did this in a manner which clearly conflicts with portions of the Bill of Rights.

Furthermore, nothing in the Act, had it been in force prior to 9/11, would have had any effect upon the events of 9/11/01. FBI agents investigating Al Qaeda suspects in the U.S., such as Coleen Rowley and Harry Samit, were stopped from actually obtaining sufficient evidence to apprehend 9/11 related suspects, not by a lack of legal authority to investigate but rather by direct orders from their superiors in the FBI itself.[248]

Samit testified unambiguously in the trial of Al Qaeda member Zacharias Moussaoui, stating that:

---

245. *United Press International*, Feb. 17, 2006, Patriot Act passes U.S. Senate hurdle, http://www.upi.com/NewsTrack/view.php?StoryID=20060217-093510-3727

246. *CBS News*, June 28, 2002, Army Lab Eyed as Anthrax Source, http://www.cbsnews.com/stories/2002/06/28/national/main513694.shtml

247. *New Scientist.com*, May 9, 2002, Anthrax bug "identical" to army strain, http://www.newscientist.com/article.ns?id=dn2265

248. *Time*, May 21, 2002, Coleen Rowley's memo to F.B.I. Director Robert Mueller, an edited version of the agent's 13 page letter, http://www.time.com/time/covers/1101020603/memo.html

...he spent four weeks warning his bosses about the radical Islamic student pilot. He said "criminal negligence" and bureaucratic resistance by FBI headquarters agents blocked "a serious opportunity to stop the 9/11 attacks."[249]

No possible legislation can overcome bureaucratic inertia, ineptness or malfeasance.

With the initial passage of the Patriot Act in 2001, the road was open to invasion of Afghanistan, which occupied territory long coveted for vital pipeline routes from Central Asia's oil and gas fields.[250] [251] [252] If the 2001 Afghan invasion was intended to capture Osama Bin Laden, named as the mastermind of the 9/11 attack, [253] the US would have had to commit significant ground forces to the invasion.

In December of 2002, when Bin Laden was pinned down and surrounded near the Pakistani border at Tora Bora, and the US knew he was there, the US again chose to delegate the ground fighting with Bin Laden's forces mainly to ill-equipped and unsupported Afghan militias. Predictably, Bin Laden easily escaped.[254] [255]

After all, Bin Laden and his Al Qaeda network were creations of the CIA during the 1979 to 1988 Soviet-Afghan War, where they served American geopolitical interests very effectively. Having escaped, Bin Laden was now an effective "bogeyman" that could be trotted out again and again in the future, whenever the government needed to frighten the American population into submission.[256] [257]

249. *Yahoo News*, March 23, 2006, Moussaoui Prosecutors Close with FBI Agent, http://news.yahoo.com/s/ap/20060323/ap_on_re_us/moussaoui

250. *BBC News World Edition*, 27 Dec., 2002, Central Asia pipeline deal signed, http://news.bbc.co.uk/2/hi/south_asia/2608713.stm

251. What Really Happened, It's All About Oil, From the 1998 Congressional Record (emphasis added to text) US Interests In The Central Asian Republics Hearing Before The Subcommittee On Asia And The Pacific Of The Committee On International Relations, House Of Representatives, One Hundred And Fifth Congress Second Session February 12, 1998, http://www.whatreallyhappened.com/oil.html

252. *Alter Net*, Feb. 28, 2002, The Enron-Cheney-Taliban Connection?, http://www.alternet.org/story/12525/

253. *9/11 review.org*, Osama Bin Asset, http://911review.org/Wiki/OsamaBinAsset.shtml

254. *The Christian Science Monitor*, March 4, 2002, How Osama bin Laden Got Away, http://www.csmonitor.com/2002/0304/p01s03-wosc.html

255. *Newsweek*, Aug. 15, 2005, Exclusive: CIA Commander: US Let bin Laden Slip Away, http://www.msnbc.msn.com/id/8853000/site/newsweek/

256. Centre for Research on Globalisation/Centre de recherche sur la mondialisation, Oct. 31, 2004, "Intelligence Asset" Osama bin Laden Supports Bush Re-election, http://www.globalresearch.ca/articles/CHO410B.html

257. *Idaho Observer*, Oct. 2001, Osama bin CIA?, http://proliberty.com/observer/20011005.htm

Even Gary Berntsen, the CIA field commander who was on the ground at Tora Bora, acknowledges this reality:

> ... in a forthcoming book, the CIA field commander for the agency's Jawbreaker team at Tora Bora, Gary Berntsen, says he and other US commanders *did* know that bin Laden was among the hundreds of fleeing Qaeda and Taliban members. Berntsen says he had definitive intelligence that bin Laden was holed up at Tora Bora — intelligence operatives had tracked him — and could have been caught. "He was there," Berntsen tells NEWSWEEK. Asked to comment on Berntsen's remarks, National Security Council spokesman Frederick Jones passed on 2004 statements from former CENTCOM commander Gen. Tommy Franks. "We don't know to this day whether Mr. bin Laden was at Tora Bora in December 2001," Franks wrote in an Oct. 19 New York Times op-ed. "Bin Laden was never within our grasp." Berntsen says Franks is "a great American. But he was not on the ground out there. I was."[258]

However, Afghanistan was not the primary goal of the corporate energy cabal — Iraq was. According to Richard Clarke, the then US terrorism czar, Bush personally ordered him to find incriminating evidence against Iraq in the aftermath of 9/11. Clarke could not do so, and did not do so. He later bitterly criticized the Bush Administration for undermining the "War on Terror" by launching an unprovoked attack on Iraq.[259] [260] [261]

The Administration next set up a special intelligence unit called the Office of Special Plans, which was staffed exclusively by PNAC-supporting neocons and which reported directly to Cheney:

> They call themselves, self-mockingly, the Cabal — a small cluster of policy advisers and analysts now based in the Pentagon's Office of Special Plans. In the past year, according to former and present Bush Administration officials, their operation, which was conceived by Paul Wolfowitz, the Deputy Secretary of Defense, has brought about a crucial change of direction in the American intelligence community. These advisers and analysts, who began their work in the days after September 11, 2001, have produced a skein of intelligence reviews that have helped to shape public opinion and American policy toward Iraq. They relied on data gathered by other intelligence agencies and also on information provided by the Iraqi National Congress, or I.N.C., the exile group headed by Ahmad Chalabi. By last fall, the operation rivaled both the C.I.A. and the Pentagon's own Defense Intelligence Agency, the D.I.A., as President Bush's main source of intelligence regarding Iraq's possible possession of weapons of mass destruction and connection with Al Qaeda. As of last week, no such weapons had been found. And although many people, within the Administration and outside it, profess confidence that something will turn up, the integrity of much of that intelligence is now in question.[262]

---

258. Ibid., *Newsweek*, # 255

259. *Slate*, March, 24, 2004, Richard Clarke KO's the Bushies, The Ex-Terrorism Official Dazzles at the 9/11 Commission Hearings, http://www.slate.com/id/2097750/

260. On the Issues.org, Against All Enemies, by Richard Clarke: on War and Peace, http://www.ontheissues.org/Archive/Against_All_Enemies_War_+_Peace.htm

261. *CBS News*, March 21, 2004, Clarke's Take on Terror, http://www.cbsnews.com/stories/2004/03/19/60minutes/main607356.shtml

False evidence justifying the invasion and occupation of Iraq was indeed manufactured and the war was launched in March of 2003. The "Downing Street Memos" became infamous in attesting to this policy of deception, because they contained:

> ...a remark attributed to Richard Dearlove (then head of British foreign intelligence service MI6) that *"the intelligence and facts were being fixed around the policy"* of removing Saddam Hussein from power, which was taken to show that US intelligence on Iraq prior to the war was deliberately falsified, rather than simply mistaken. As this issue began to be covered by American media, two other main allegations stemming from the memo arose: that the UN weapons inspection process was manipulated to provide a legal pretext for the war, and that pre-war air strikes were deliberately ramped up in order to soften Iraqi infrastructure in preparation for war, prior to the October congressional vote permitting the invasion.[263]

It is interesting to observe how eerily similar this methodology is to that employed by the US Geological Survey, discussed in Chapter 7, to estimate remaining oil reserves: invent the facts.

Professor Noam Chomsky has described the rigid guidelines defining mass-marketed "reality" concerning what can and cannot be said in the media on the subject of why the US invaded Iraq:

> [We] are under a rigid doctrine in the West, a religious fanaticism that says we must believe that the United States would have invaded Iraq even if its main product was lettuce and pickles, and the oil resources of the world were in Central Africa. Anyone who doesn't believe that is condemned as a conspiracy theorist, a Marxist, a madman, or something. Well, you know, if you have three gray cells functioning, you know that that's perfect nonsense. The US invaded Iraq because it has enormous oil resources, mostly untapped, and it's right in the heart of the world's energy system. Which means that if the US manages to control Iraq, it extends enormously its strategic power, what Zbigniew Brzezinski calls its critical leverage over Europe and Asia. Yeah, that's a major reason for controlling the oil resources — it gives you strategic power. Even if you're on renewable energy you want to do that. So that's the reason for invading Iraq, the fundamental reason.[264]

The Constitution states that when the United States ratifies a treaty, that treaty becomes the law of the nation. The United States was signatory to the Hague Convention of 1954 and Geneva Convention of 1949. These Conventions constrain an occupying power from seizing control of the occupied nation's economic resources for itself. However, once the US occupied Iraq, Proconsul Paul Bremer issued one hundred "Transitional Administrative Laws" which could not

---

262. *The New Yorker*, May 12, 2003, Selective Intelligence, Donald Rumsfeld Has His Own Special Sources. Are They Reliable?, http://www.newyorker.com/fact/content/?030512fa_fact

263. The Sunday Times Britain, May 1, 2005, Secret and Strictly Personal UK Eyes Only, http://www.timesonline.co.uk/article/0,,2087-1593607,00.html

264. *Alter Net*, Jan., 14, 2006, Chomsky: "There is no War on Terror", http://www.alternet.org/story/30487/

be repealed or amended for years to come by future Iraqi governments.[265] These laws give effective control of Iraq's economy, and most importantly its oil, to US- and "coalition"-based multinational firms.[266] "Mission accomplished" indeed!

In summary, a regime that is wholly beholden to multinational energy companies was installed into power in the United States in late 2000 and was reinstalled in 2004. Within their first year, they seized the advantage of their galvanizing "new Pearl Harbor" event, using that tragedy to usher in an era of "war that will not end in our lifetimes..."[267] and to eviscerate the US Consti- tution and its system of law. America's courts are now increasingly packed with their appointees; the rule of law, along with judicial independence, is fast fading.

The US, and the world, have been hijacked. As this control is consolidated, an "unholy union" of Corporation with Church and State is emerging as the dominant organizing paradigm of the late modern world.

265. *Global Issues.org*, July 4, 2004, Iraq: War and the Aftermath, Handover of Power to Iraqis, http://www.globalissues.org/Geopolitics/MiddleEast/Iraq/PostWar/ Handover.asp?p=1

266. Cash From Chaos 1: corruption and neo-liberal rule in Iraq, Dave Whyte, University of Stirling, unpublished manuscript, http://www.dass.stir.ac.uk/staff/d-whyte/docu- ments/CashFromChaos1.pdf

267. *9/11 Truth.org*, Jan. 18, 2005, Crossing the Rubicion: Simplifying the Case Against Dick Cheney, http://www.911truth.org/article.php?story=20050119084227272

CHAPTER 9. UNHOLY UNION: CHURCH, CORPORATION, AND STATE

> *If you think of yourselves as helpless and ineffectual, it is certain that you will create a despotic government to be your master. The wise despot, therefore, maintains among his subjects a popular sense that they are helpless and ineffectual.* — Frank Herbert[268]

DISPENSATIONALISM, RECONSTRUCTIONISM, AND THE CHURCH STATE INC.

The corporatists have established a marriage of convenience with the burgeoning religious right. America was founded by both plutocrats (for example, Jamestown was a corporate sponsored for profit venture) and theocrats (for example, the Massachusetts Bay Colony, an intolerant theocracy). This unholy union of corporatism and religion represents the consummation of a North American political incest. Conservative, politically active Christians, Jews and others, who will reliably vote and think as instructed by their leaders, and the corporate elite who rely upon those easily-manipulated shock troops, work against real democracy.

Today's Christian fundamentalism has its origins in the 16th century doctrine of Calvinism. Calvinism has always maintained that wealth and power are signs of God's favor, and suggests that those who possess these attributes are to be obeyed and respected by those who do not. Corporatism could not have gained nor maintained political control without the religious mass base being swayed by such ideas.

---

268. Quoteland.com: Frank Herbert, http://www.quoteland.com/author.asp?AUTHOR_ID=126

Politically active fundamentalist Christians are primarily drawn from a school of religious thought that originated in mid-19th century England called Dispensationalism. Many people now assume that Dispensationalist theology is, and always has been, orthodox Christian theology, but nothing could be farther from the truth. Among other things, Dispensationalists believe that we have entered into the final period of history during which the Bible's prophecies will be carried out literally, and the world as we know it will end. Dispensationalism's founder, John Nelson Darby, in the early 1840s, invented the notion of the pre-tribulational rapture of the Church. Darby claimed that seven years before Christ returns to Earth, all believing Christians (apparently this applies only to Dispensationalists) will be suddenly "raptured" that is, they will be made to disappear forever from Earth as they will have been transported directly to Heaven to spend eternity with God.

Televangelists such as Falwell, Robertson, LaHaye, and Dobson have spread this belief system worldwide.

Dispensationalism would have been classified as a major doctrinal heresy anytime before the Protestant Reformation; after the Reformation, it would simply have been classified as a fringe sect. But recently Dispensationalist dogma has mutated further. Its emphasis has come to focus ever more intently upon the end times and the "rapture." Jesus has come to be portrayed as a bloody avenger.

This cult-like religious doctrine emerged when England was in the throes of the Industrial Revolution between the years of 1830-1870. The population was being driven off the farms en masse and herded into factories — and the social disruption this entailed was horrific. This was a time of mass poverty, social dislocation, and general dehumanization. Religious institutions such as the Church of England were identified with the capitalist factory owner class and were widely perceived to be unresponsive to the needs of the working class.

Because the industrial era was clearly a radical break with the agrarian past, a religion for a new era was needed. And the tenets of Dispensationalism offered the people hope of a "new dispensation from God" in the form of the end of the world.

The Dispensationalist dogma instructs those who are on the lower economic end of the corporate dominated capitalistic world system to grit their teeth and endure their demeaning roles in it while waiting for those whom they see as being responsible to get their well-deserved comeuppance from God. It thereby renders these industrial workers politically non-threatening to the corporate owners.

Further, the corporatists who are creating the new world reality are not held accountable for it. Rather, the principal opponents of the corporatists, including liberals, labor unions, and vaguely defined "secularists," are blamed for the travails of the working class.

In the US, Dispensationalism's 30 to 40 million or so adherents have become a key constituency of the Republican Party and comprise much of the voting base for the military-industrial-media complex dominated by global hydrocarbon multinationals..

The eschatology (doctrinal beliefs about the end of the world) of Dispensationalists is strongly focused upon one aspect of the coming "end times": their own much-desired "rapture." Secondarily, it is focused upon the subsequent destruction of almost everyone else. Dispensationalists believe that as part of this ending of the world process, the various covenants made by God with Israel will be fulfilled in the near future. Fulfilling these covenants will require a global war centered upon the Middle East, culminating in Israel's and its allies' victory over the forces of evil at the Battle of Armageddon in central Israel.

Thus Dispensationalists are strongly pre-disposed to desire war in the Middle-East — something which they believe is pre-ordained by God anyway — as soon as possible, and as intensively as possible. To them, the US is interested in Middle-Eastern nations not to gain control of two-thirds of the planet's remaining oil, but to bring on the prophecy. Even wishing to stop these wars is seen by Dispensationalists as opposing God's will.

At the same time, Dispensationalists welcome ecological disaster as a sign of the onset of the "end times," their personal rapture, and the coming of the Kingdom of God on Earth.

Thus Dispensationalists engage in political actions which advance the interests of the corporate juggernaut rather than their own personal material interests. What counts for these "believers" is advancing their perceived spiritual interests, even if it means rejecting the notion of being stewards of the earth.

Thus, both the corporate agenda of maximal short-term profit and the Dispensationalist agenda of rapture politics lead to the same outcome: destruction.

Then there are the Reconstructionists. They believe that Christ will return when the entire world is a Reconstructionist theocracy. Reconstructionism is unabashedly pro-free market capitalism, meaning that it is rabidly corporatist. They find support for unfettered market capitalism in the Bible.[269]

The merger of theological fundamentalism with free-market fundamentalism is complete. The outline of a seamless religiously mandated corporatist dictatorship begins to emerge in their dogma. Of course, support for the opposite can also be found in the Bible.

Kevin Phillips writes in *American Theocracy*:

> The most intense believers split into two principle camps. The first, so called Dispensationalists, who interpret current events such as the tsunamis, oil spikes, and wars as confirming that end times are at hand, usually don't worry about energy policy. Indeed, they cite a biblical verse mentioning costly wheat, barley, and oil as predictions of shortages of food and fuel....Reconstructionists, by contrast, believe

269. *Encyclopædia Britannica*, 2006, fundamentalism, Encyclopædia Britannica Premium Service, http://www.britannica.com/eb/article-252655.

that the world must be made over theocratically, along biblical lines, before Christ will return. Neither faction has fossil fuels, climate deterioration, or the energy efficiency of the US manufacturing sector on its agenda. Both camps deplore the efforts of geologists and climatologists to sway voters and policy-makers through Hubbert's peak analyses and scientific interpretations of global warming data. Their biblically viewed world is at most ten thousand years old, not the millions of years established by scientists, whose insistence on this longer timeframe is said to usurp God's prerogative. In considering stem-cell research or Iraq-as-Babylon, depleting oil, or melting polar ice caps, the thought processes of such true believers have at best limited openness to any national secular dialogue.[270]

Humanity's religions, including Christianity, Islam, Judaism, Buddhism, and others, were intended to embody noble aspirations. However, political agendas masquerading as religion are used to capture the imagination of the public and substitute noxious ideals for life-affirming values.

Adam Smith, author of the canonical capitalist "Bible" *Wealth of Nations*, written in the intellectually, politically, and economically transformative year of 1776, never envisioned an oligopolistic market consisting of global multinational firms which effectively control the planet's governments, as opposed to a market consisting of smaller firms which are firmly controlled and regulated by national governments. Smith assumed that a market would consist of these smaller firms which were themselves firmly embedded in a society's systems of ethics and values. What we have today is no Smithian free market. It is dramatically different to what he envisioned.

David Ricardo was a British economist who is best known for his 1817 book *Principles of Political Economy and Taxation*, in which he develops his now canonical theory of comparative advantage. Ricardo demonstrated logically and mathematically that whenever two or more nations trade with one another, *assuming that the trading is free and fair*, the result is always greater production and distribution (via trade) of wealth than would occur if each of the countries produced all of its goods domestically. This is true even where one country can produce every good than another country produces, *and* can produce all of those goods more efficiently, because every firm in each country produces those goods which it is best at producing with maximum efficiency. Thus overall efficiency of production is maximized. However, in formulating this argument, Ricardo assumed that each nation's productive resources in the form of land (land itself, along with factories), labor, and capital, known to classical economics as the three factors of production, would remain within a nation's control. These factors of production could be realigned within a nation's borders. However, *they could not be exported from one nation to another nation.* Yet modern globalized multinational capitalism is predicated upon just this ability to export factors of pro-

---

270. Phillips, Kevin, American Theocracy The Peril and Politics of Radical Religion, Oil, and Borrowed Money, Viking-Penguin, New York, NY, USA, 2006, pps 66-67

duction at whim in order to maximize short-term profits. This constitutes another massive "heresy" with respect to orthodox capitalist doctrine.

Thus, the corporatists who are closely allied with Dispensationalists are promoting a skewed vision of capitalism with respect to the ideas presented by Adam Smith and David Ricardo.

Political reformer David Sirota observed in the *San Francisco Chronicle* that:

> Amid all the consultant-packaged rhetoric about America being the "greatest democracy in the world," it often seems impossible to figure out exactly who controls our government. But every now and then, the public gets a fleeting glimpse into who is really running the show. We get to see how there no longer is a boundary between Big Business and government, and how our politicians are wholly owned subsidiaries of Corporate America. We get to see, in short, exactly how our government has been the victim of a hostile takeover. Last month, in three little-noticed stories buried in the business press, the hostile takeover was on full display. The first story was a tiny one buried on the inside pages of the Wall Street Journal about how the US Treasury Department worked hand in hand with IBM to kill bipartisan pension legislation in 2003. The bill would have outlawed pension schemes employed by IBM and other big companies that give workers less than they were originally promised. The report noted that..., "a Treasury official disclosed nonpublic information to IBM and failed to report expenses paid by a lobbyist for a pension-industry trade group" — all the while allowing the company to circulate documents on Capitol Hill claiming the US Treasury officially was working with IBM to kill the legislation. Clearly, the behavior ran afoul of the lobbying laws supposedly creating a boundary between business and government. But as the Journal went on to note, "The Justice Department didn't pursue criminal or civil charges in the matters because they didn't meet the agency's 'prosecutorial threshold.' The legislation was ultimately killed. In effect, a major federal agency — in this case the Treasury Department — was the victim of the hostile takeover, serving as an arm of Corporate America, rather than a regulator.[271]

Bill Moyers is currently the President of the Schumann Center for Media and Democracy. He is a former PBS commentator, as well as being a former Special Assistant for President Lyndon B. Johnson, whom he also served as a key advisor, speechwriter, and informal White House Chief of Staff.[272] Moyers too, is well aware of the ongoing hostile takeover of American democracy:

> For a quarter of a century now a ferocious campaign has been conducted to dismantle the political institutions, the legal and statutory canons, and the intellectual, cultural, and religious frameworks that sustained America's social contract. The corporate, political, and religious right converged in a movement that for a long time only they understood because they are its advocates, its architects, and its beneficiaries. Their economic strategy was to cut workforces and wages, scour the

---

271. *San Francisco Chronicle*, May, 1, 2006, Fighting the Hostile Takeover, http://www.sfgate.com/cgi-bin/article.cgi?file=/chronicle/archive/2006/05/01/EDGK5IFLNP1.DTL

272. MBC: Museum of Broadcast Communications, Moyers, Bill, US Broadcast Journalist, http://www.museum.tv/archives/etv/M/htmlM/moyesrbill/moyersbill.htm

globe for even cheaper labor, and relieve investors of any responsibility for the cost of society. On the weekend before President Bush's second inauguration, *The New York Times* described how his first round of tax cuts had already brought our tax code closer to a system under which income on wealth would not be taxed at all and public expenditures would be raised exclusively from salaries and wages. Their political strategy was to neutralize the independent media, create their own propaganda machine with a partisan press, and flood their coffers with rivers of money from those who stand to benefit from the transfer of public resources to elite control. Along the way they would burden the nation with structural deficits that will last until our children's children are ready to retire, systematically stripping government of its capacity, over time, to do little more than wage war and reward privilege. Their religious strategy was to fuse ideology and theology into a worldview freed of the impurities of compromise, claim for America the status of God's favored among nations (and therefore beyond political critique or challenge), and demonize their opponents as ungodly and immoral. At the intersection of these three strategies was money: Big Money. They found a deep flaw in our political system and zeroed in on it.[273]

CORPORATE TAKEOVER

The merger of corporate governance with the governing apparatus of the various nation-states, legitimated by religion and or ideology, is rapidly nearing full consummation. The methodology utilized to accomplish this end differs between major nations such as India, Japan, Britain, US, the Russian Federation and the People's Republic of China;[274] [275] [276] [277] [278] [279] [280] however, the end results are functionally equivalent. The result is a political system characterized by the merging of state and corporation with the corporate elite, exercising

273. *Tom Paine.com*, March 22, 2006, A Time for Heresy, http://www.tompaine.com/articles/2006/03/22/a_time_for_heresy.php

274. *The Christian Science Monitor*, Dec. 28, 2005, Kremlin reasserts control of oil, gas, http://www.csmonitor.com/2005/1228/p01s01-woeu.html

275. *Alter Net*, May 10, 2005, The Intensifying Global Struggle for Energy, http://www.alternet.org/envirohealth/21969/

276. *Index Research*, Nov. 18, 2005, Siberian Shadowlands-Part 2, Part II: Corporate Power: The New Tsar, http://indexresearch.blogspot.com/2005/11/siberian-shadowlands-part-2.html

277. *Power and Interest News Report*, Sept. 2, 2005, Economic Brief: China's Energy Acquisition,http://www.pinr.com/report.php?ac=view_report&report_id=359&language_id=1

278. Public Citizen: Protecting Health, Safety, and Democracy: Corporate Control, http://www.citizen.org/cmep/corporatecon/

279. Rand Corporation Commentary (Originally published in *New York Times*, Aug. 13, 2001) China's Capitalists Join the Party, http://www.rand.org/commentary/081301NYT.html

280. Oil Companies.net, ExxonMobil Reports Annual Profits of $25 Billion, http://www.oilcompanies.net/

undivided power which is legitimated by the organs of the state. Nationalist ideology and religion are controlled by these power elites to further their self-serving goals, while delegitimating any and all who oppose them.

A class-based society ruled by wealthy corporate elites has emerged, and it is strongly legitimated by the power structure of the society. The police powers of the state are brought to bear against opponents. Elections are effectively uncontested, or else are fraudulent; only candidates beholden to the power elite are permitted to win. The media is employed in the service of corporate interests.

Fear is the essential organizing principle of a fascist society. Fear renders citizens helpless and dependent upon government for safety. Fear compels their obedience. And the need for safety — for "homeland security" — overrides all else.

Dr. Paul Craig Roberts, a former Undersecretary of the Treasury in the Reagan Administration, is a life-long Republican and a conservative of the old school which believes in small government, fiscal responsibility, rule of the people, and adherence to the US Constitution. He has parsed the dense text of the Patriot Act renewal legislation, which despite minor amending was renewed and made permanent by Congress. He notes that:

> A provision in the "PATRIOT Act" creates a new federal police force with the power to violate the Bill of Rights. You might think that this cannot be true, as you have not read about it in newspapers or heard it discussed by talking heads on TV....Sec. 605 reads: *"There is hereby created and established a permanent police force, to be known as the 'United States Secret Service Uniformed Division.'"* This new federal police force is "subject to the supervision of the Secretary of Homeland Security." The new police are empowered to "make arrests without warrant for any offense against the United States committed in their presence, or for any felony cognizable under the laws of the United States if they have reasonable grounds to believe that the person to be arrested has committed or is committing such felony." The new police are assigned a variety of jurisdictions, including "an event designated under section 3056(e) of title 18 as a special event of national significance" (SENS). "A special event of national significance" is neither defined nor does it require the presence of a "protected person" such as the president in order to trigger it. Thus, the administration, and perhaps the police themselves, can place the SENS designation on any event. Once a SENS designation is placed on an event, the new federal police are empowered to keep out and arrest people at their discretion. The language conveys enormous discretionary and arbitrary powers. What is "an offense against the United States"? What are "reasonable grounds"? You can bet the Alito/Roberts court will rule that it is whatever the executive branch says.[281]

---

281. *Anti-War.com*, Jan. 24, 2006, Unfathomed Dangers in PATRIOT Act Reauthorization, http://www.antiwar.com/roberts/?articleid=8434 Boston Globe, April 30, 2006, Bush Challenges Hundreds of Laws, President Cites Powers of His Office, http://www.boston.com/news/nation/washington/articles/2006/04/30/bush_challenges_hundreds_of_laws/?page=1

This new and improved system of corporatism appropriating the power and symbols of the national state, legitimated by intolerant religion, might plausibly be called neo-fascism, "neo" in order to distinguish it from its cruder predecessor. However, the same system can simply be called "corporatism." It's just a question of where you draw the line.

The United States Secret Service Uniformed Division, acting as a national police force, effectively unconstrained by any law, probably will not flaunt itself quite so crudely and blatantly as did their brown- and black-shirted predecessors. Unfortunately, it simply does not matter. Once a national police force emerges that is answerable only to the leader and is not constrained by law, the game is over: Democracy, rule of the people, by the people, and for the people, is effectively dead.

George W. Bush has claimed the right to annul any law passed by Congress, as well as the power to override any decision of the United States Supreme Court, as reported by the *Boston Globe*:

> President Bush has quietly claimed the authority to disobey more than 750 laws enacted since he took office, asserting that he has the power to set aside any statute passed by Congress when it conflicts with his interpretation of the Constitution. Among the laws Bush said he can ignore are military rules and regulations, affirmative-action provisions, requirements that Congress be told about immigration services problems, "whistle-blower" protections for nuclear regulatory officials, and safeguards against political interference in federally funded research....David Golove, a New York University law professor who specializes in executive-power issues, said Bush has cast a cloud over "the whole idea that there is a rule of law," because no one can be certain of which laws Bush thinks are valid and which he thinks he can ignore. "Where you have a president who is willing to declare vast quantities of the legislation that is passed during his term unconstitutional, it implies that he also thinks a very significant amount of the other laws that were already on the books before he became president are also unconstitutional," Golove said...Bush has also challenged statutes in which Congress gave certain executive branch officials the power to act independently of the president. The Supreme Court has repeatedly endorsed the power of Congress to make such arrangements. For example, the court has upheld laws creating special prosecutors free of Justice Department oversight and insulating the board of the Federal Trade Commission from political interference. Nonetheless, Bush has said in his signing statements that the Constitution lets him control any executive official, no matter what a statute passed by Congress might say.[282]

What is the definition of tyranny? Of a dictatorship? The only possible remedy for this behavior is impeachment. Any action short of impeachment means that the precedent of unlimited presidential power has been successfully set and has been accepted.

---

282. *Boston Globe*, April 30, 2006, Bush Challenges Hundreds of Laws, President Cites Powers of His Office, http://www.boston.com/news/nation/washington/articles/2006/04/30/bush_challenges_hundreds_of_laws/?page=1

Former *Newsweek* and UPI correspondent Robert Parry broke many of the stories concerning the abuse of presidential powers during the 1980s Iran Contra scandals. He has remained an astute observer of the unraveling of the political system before concentrated corporate power during the intervening decades. Parry writes:

> Every American school child is taught that in the United States, people have "unalienable rights," heralded by the Declaration of Independence and enshrined in the US Constitution and Bill of Rights. Supposedly, these liberties can't be taken away, but they are now gone. Today, Americans have rights only at George W. Bush's forbearance. Under new legal theories — propounded by Supreme Court nominee Samuel Alito and other right-wing jurists — Bush effectively holds all power over all Americans. He can spy on anyone he wants without a court order; he can throw anyone into jail without due process; he can order torture or other degrading treatment regardless of a new law enacted a month ago; he can launch wars without congressional approval; he can assassinate people whom he deems to be the enemy even if he knows that innocent people, including children, will die, too. Under the new theories, Bush can act both domestically and internationally. His powers know no bounds and no boundaries. Bush has made this radical change in the American political system by combining what his legal advisers call the "plenary" — or unlimited — powers of the Commander in Chief with the concept of a "unitary executive" in control of all laws and regulations. Yet, maybe because Bush's assertion of power is so extraordinary, almost no one dares connect the dots. After a 230-year run, the "unalienable rights" — as enunciated by Thomas Jefferson, James Madison and the Founding Fathers — are history.[283]

An all powerful Presidency is being created. Bush's assertion of essentially unlimited powers under the unconstitutional doctrine of the "Unitary Executive" which includes the "right" to disregard any law because it is "wartime" marks the definitive end of the American republic.

Simultaneously, growing multitudes of enraged victims of Western war-making in Iraq, Afghanistan, Pakistan, and across the planet, became available to swell the ranks of terrorists focused on obtaining revenge against the US and its allies. As did growing cohorts of angry civilians opposing the various regimes scattered across the Middle East which were propped up by US and Western military force, despite all talk of "democratizing" the region.

In an interview in *Cigar Aficionado* General Tommy Franks commented as follows:

> What is the worst thing that can happen in our country? The worst thing that can happen is, perhaps — and this is my personal opinion — two steps. The first step would be a nexus between weapons of mass destruction of any variety. It could be chemical, it could be biological, it could be some nuclear device; and terrorism. Terrorists or any human being who is committed to the proposition of terror, try to just create casualties, not for the purpose of annihilation, but to terrify a population. We see it in the Middle East today, in order to change the mannerisms, the behavior, the

---

283. *Consortium News.com*, Jan. 24, 2006, The End of "Unalienable Rights", http://www.consortiumnews.com/2006/012406.html

sociology and, ultimately, the anthropology of a society. That goes to step number two, which is that the western world, the free world, loses what it cherishes most, and that is freedom and liberty we've seen for a couple of hundred years in this grand experiment that we call democracy. Now, in a practical sense, what does that mean? It means the potential of a weapon of mass destruction and a terrorist, massive casualty-producing event somewhere in the western world — it may be in the United States of America — that causes our population to question our own Constitution and to begin to militarize our country in order to avoid a repeat of another mass-casualty-producing event. Which, in fact, then begins to potentially unravel the fabric of our Constitution. Two steps: very, very important.[284]

And so the "War on Terror" creates fresh legions of terrorists, who by attacking the vital energy infrastructure create such vast disruption as to bring about the end of what remains of constitutional government.

An unanticipated strong and popularly supported guerrilla resistance in Iraq, generated in major part by US blundering, has prevented the corporatist neo-cons thus far from striking at Iran. This sequence of wars is very clearly a continuous, ongoing, global war for control over oil supplies.[285] [286]

Recall Figure 9 from Chapter 3:

Figure 1. The Oil Patch.
Source: Early Warning Report (See Chapter 3).

284. *Cigar Aficionado Magazine*, Dec. 1, 2003, General Tommy Franks: An Exclusive Interview With America's top General in the War on Terrorism, http://www.cigaraficionado.com/Cigar/CA_Profiles/People_Profile/0,2540,201,00.html

285. *Oil Companies.net*, The New US-British Imperialism Part 2, http://www.oilcompanies.net/oil2.htm

286. *Mercury News*, Jan. 30, 2005, Nearly Half of Iraqis Support Attacks on US Troops, http://www.mercurynews.com/mld/mercurynews/news/politics/13750080.htm?template=contentModules/printstory.jsp

It is indeed all about oil: with only four percent of the planet's population, the US consumes about one quarter of the world's oil. When you are a hammer, everything looks like a nail.

Still, other great powers, like Russia, which borders upon Iran, and most significantly China, which lacks Russia's considerable domestic oil reserves, cannot afford to allow Iran's resources to fall under American control.

In summary, the world's governments — and most particularly the US government — will respond one way or another to the challenges posed by the many interrelated crises now bearing rapidly down upon us. However, the corporate entities which are actually causing these crises have the final word in how the world's governments respond, and the short-term profit motive argues against steering toward effective, real solutions; more frequently, they will actually implement reactionary policies which shall make the situation far worse.

Government in this day and age is commonly supposed to be the tool by which we solve common problems — particularly those which concern survival. However, government now serves the interest of disembodied corporate "persons." David Goodstein is a physics professor at Caltech. He concludes his book *Out of Gas: The End of the Age of Oil* by commenting:

> We can envision a future in which we live entirely on nuclear energy and solar energy as it arrives from the Sun. That would not require a reversion to an eighteenth-century lifestyle and a concomitant drastic reduction of the human population of the earth. Instead it would be based upon a sophisticated technology which converts sunlight and nuclear energy efficiently into electricity for stationary uses, and produces hydrogen fuel or charges advanced batteries for mobile uses. That would leave the carbon in the ground, or at least unburned, as a source for the petrochemicals that are also an indispensable feature of our way of life. And it might be no more difficult to accomplish than putting people on the moon. Unfortunately, our present national and international leadership is reluctant even to acknowledge that there is a problem. The crisis [peak oil] will occur and it will be painful. The best we can realistically hope for is that when it happens, it will serve as a wake-up call and will not so badly undermine our strength that we will be unable to take the giant steps that are needed. For now I stand by the warning I made in the first paragraph of this book's introduction: Civilization as we know it will come to an end sometime in the next century unless we can find a way to live without fossil fuels.[287]

Professor Goodstein notes the failure of national and international leadership to deal with the imminent existential crisis posed by peak oil, but he does not address the related crises of global warming, climate change, famine, war and so on.

---

287. Goodstein, David, *Out of Gas: The End of the Age of Oil*, W.W. Norton & Company, New York, NY, 2004, pp. 123.

Chapter 10. The Big Picture

The true civilization is where every man gives to every other
every right that he claims for himself. — Robert Ingersoll[288]

View from the Peak

The global political economy is almost wholly predicated upon the assumption that limitless quantities of cheap hydrocarbon energy are available, and that whatever humans do will not seriously affect the planet's life support systems. And that if need be, some substitute for cheap hydrocarbons can always be found. All of these assumptions are false.

The 21st century global political economy can be considered to be a large complex-adaptive system which is nested within an even larger adaptive system upon which it depends for its very existence: nature. The outputs of the human system are driving the surrounding natural system from its long-term climate equilibrium. This sets off a series of inter-related changes, all of which tend toward fuel shortages, food shortages, water shortages and health problems so severe that a desperate struggle for survival may ensue.

Humanity is now entering into a period when major changes must be made; it is an opportunity for a total memetic replacement by a more adaptive understanding of nature and our relationship to it. Let us hope we evolve a global system founded upon sharing and caring — not because this is "good" but because it facilitates survival itself. Sharing, caring, understanding that one is a part of a nested system of systems, emerge as survival values.

---

288. The Quotations Page: Robert Ingersoll, http://www.quotationspage.com/quote/28996.html

BEWARE "MAGIC BULLETS"

Awareness of the reality of global warming has now reached almost every-where — although it is perhaps not openly admitted. (Imagine the worldwide panic if it were.)

A critical mass of public support is building in favor of doing something decisive to redress the problem. However, it is hard to see which countries (and voters) and which corporations (and owners) will be the first to say, "I'll stop being greedy and leave the goodies there, and I'll trust all the rest of you to do the same."

Think tanks like the Hoover Institute are proposing miracle work-arounds like pumping tens of millions of tons of sulfate particles into the stratosphere to filter out the near ultraviolet component of atmospheric sunlight, the compo-nents that account for, among other things, sunburn and skin cancer. Humanity could then continue to burn hydrocarbons without limit and without any envi-ronmental constraint. We could have our cake and eat it too.

Regrettably, they seem to overlook several complicating facts. For one, many insects including honey bees depend upon UV radiation to select flowers and other plants for feeding upon.[289] If the flower is not illuminated by UV, they will reject it for feeding.[290] Substantially decreasing ambient UV levels could destroy the pollinators upon which agriculture depends. Knocking out much of this UV radiation could severely disrupt the ability of vast numbers of species — possibly millions of species — to navigate, feed, and reproduce.

In any event, global warming is not the primary problem. Rather, it is a sign of something much deeper.

> The relative indifference to the environment springs, I believe, from deep within human nature. The human brain evidently evolved to commit itself emotionally only to a small piece of geography, a limited band of kinsmen, and two or three genera-tions into the future. To look neither far ahead nor far afield is elemental in a Dar-winian sense. We are innately inclined to ignore any distant possibility not yet requiring examination. It is, people say, just good common sense. Why do they think in this shortsighted way? The reason is simple: It is a hardwired part of our Paleolithic heritage. For hundreds of millennia, those who worked for short term gain within a small circle of relatives and friends lived longer and left more offspring — even when their collective striving caused their chiefdoms and empires to crum-ble around them. That long view that might have saved their distant descendants required a vision and extended altruism difficult to marshal. The great dilemma of

289. Ultraviolet Reflectance Characteristics in Flowers of Crucifers, *American Journal of Botany*, Aug. 1972, 59(7): 706-713, http://links.jstor.org/sici?sici=0002-9122%28197208%2959%3A7%3C706%3AURCIFO%3E2.0.CO%3B2-L&size=LARGE

290. A simple field method for manipulating ultraviolet reflectance of flowers, *Canadian Journal of Botany*, Volume 80, Number 12, Dec. 2002, pp. 1325-1328(4), http://www.ingentaconnect.com/content/nrc/cjb/2002/00000080/00000012/art00010

environmental reasoning stems from this conflict between short-term and long-term values. To select values for the near future of one's own tribe or country is relatively easy. To select values for the distant future of the whole planet also is relatively easy — in theory at least. To combine the two visions to create a universal environmental ethic is, on the other hand, very difficult. But combine them we must, because a universal ethic is the only guide by which humanity and the rest of life can be safely conducted through the bottleneck into which our species has foolishly blundered.[291]

At present, humanity's choices are not made on the basis of the long view but rather are almost entirely based upon very short-term considerations. Meanwhile, schemes ranging from stratospheric blocking to ethanol seem to be promoted at least in part to prevent more appropriate measures from being taken. And meanwhile, corporations who would provide the sulfate or the corn have their eyes on profits that would rival those of Big Oil.

Experts are also proposing that nuclear power plants be used to "cook" oil shale deposits into oil. Nuclear power suffers from severe fuel constraints of its own. If any new nuclear plants are to be built, it would be best to use to use them for electricity production as part of a program of moving rapidly to electric rail for long haul transportation. This would allow the quick phase-out of energy-inefficient long haul trucking fleets, long distance bussing, and oil burning locomotives, at considerable hydrocarbon fuel savings.

Humans can only learn fundamentally new behaviors through the mechanism of crisis, when old behaviors fail, leading to crisis. Survival then requires innovation. Once the lessons are learned and incorporated, civilization will be able to reconstitute itself, freed from its previous delusions about reality.

Presumably, the corporations are readying a new strategy that involves admitting the existence of a massive global warming problem. After building public anxiety to the level of sheer terror, their "magic bullet" solution will suddenly be unveiled. This will be packaged to make continued, even accelerated pollution, sound like a good thing: increased CO_2 pollution in the atmosphere allows plants to grow more vigorously, allowing for billions more people to be fed.

Whatever the form of the next several corporatist "magic bullets," an alert resident of planet Earth should ask who is first in line to profit from the new technology (whether it is effective in solving any problem or not — as long as it is widely adopted) and see through the sales pitch.

291. Wilson, Edward O., *The Future of Life*, Alfred A Knopf Publishers, New York, NY., 2004, pp. 40.

TECHNO-OPTIMISM AND REALITY

Optimists sometimes cite counterarguments that suggest that market pricing signals will facilitate the development of technology which will resolve any and every problem or shortage.

Armory Lovins and the Rocky Mountain Institute released a study entitled *Winning the Oil Endgame, Innovations for Profits, Jobs, and Security.*[292] This documented a coherent plan for a transition from the hydrocarbon-based economy to a renewable energy based one between 2005 and 2025. Scrutiny of this study reveals a fatal flaw in its underlying assumptions. The report states:

> We uncritically adopt the EIA's [US Energy Information Agency] official projection that in 2025, 20% more Americans — 348 million people — will use 40% more energy and 44% more oil than 289 million Americans used in 2002. This extrapolation reflects a far from deprived future: a 96% higher GDP, 97% bigger personal disposable income, 22% larger labor force, 80% higher industrial output, 81% more freight trucking, 41% more commercial floorspace, and 25% more, and 6% bigger houses. Each person will drive 33% and fly 48% more miles. Total light vehicle miles rise 67%...[293]

The Rocky Mountain Institute study goes on to note that electricity costs are projected to decrease by 4% by 2025. They fail to indicate the basis for such an unlikely conjecture.

The US government's projections of future supply of oil are not grounded in fact. Therefore, the Rocky Mountain Institute study's core assumption of ever rising oil production is false, so the entire study is essentially worthless.

Perhaps the best known exponent of this techno-optimistic approach was the late Nobel Prize winning economist Julian Simon. Simon argued that since the capacity of the human mind to solve problems was limitless, it followed that there were no material limits upon humanity. In his 1982 book, *The Ultimate Resource*,[294] Simon actually asserted:

> Our energy supply is non-finite, and oil is an important example . . . the number of oil wells that will eventually produce oil, and in what quantities, is not known or measurable at present and probably never will be, and hence is not meaningfully finite.[295]

Simon's core argument is simple: because the capabilities of the human mind are effectively unlimited, human ingenuity will figure out how to obtain

292. Lovins, Armory, B., et al., *Winning the Oil Endgame, Innovation for Profits, Jobs, and Security*, Rocky Mountain Institute, Snowmass, CO, 2004, http://www.oilendgame.com
293. Ibid., Lovins, # 338, pps. 37-38.
294. Simon, Julian, *The Ultimate Resource*, Princeton University Press, Princeton, NJ, 1996.
295. Minnesotans for Sustainability, Jan. 1982, The Ultimate Resource by Julian Simon, Review by Herman E. Daly, http://www.mnforsustain.org/daly_h_simon_ultimate_resource_review.htm

whatever quantities of energy humanity requires at any given time. Corollary to this is the assumption that markets are efficient and will, through the mechanism of price signals, generate the investment needed to produce more energy more cheaply. Rising costs of energy will attract investment to find new sources of energy and/or in research and development of novel energy sources. And so supply always at least equals demand, *ad infinitum*.

Unfortunately this seems to be a matter of faith rather than fact. There are only so many hydrocarbon energy resources within the earth, and much of what remains would cost too much in energy to extract. Technological gains in efficiency are themselves constrained by the laws of thermodynamics.

Still, some kind of radical breakthrough must be hoped for, and as far as ideas are concerned, the ingenuity does seem endless. It's the implementation that is most difficult, because these schemes all would require enormous degrees of sharing and trust. The single most difficult challenge for any energy production and distribution network is storage. There is no effective way to store large amounts of power. Additionally, the greatest disadvantage of solar and wind power systems is that they cannot operate continuously — the sun does not always shine and the wind does not always blow. Global distribution of renewable power addresses all of these problems. Power does not need to be stored — at least not in great quantities. Wind power from the night side of the world can be sent to the day side where it is most needed. Excess solar power gathered in the summer hemisphere can be sent to augment the weaker intake in the winter hemisphere. Wind power from areas where the wind is blowing can take up the slack for those which are becalmed.

One high-tech "miracle" solution that has been proposed is to place microwave reflectors in geosynchronous orbit to facilitate global power sharing. The high cost of access to space was a major deterrent, but new private-sector ventures such as Virgin Galactic[296] and the builder of its rocket engines SpaceDev[297] are poised to make relatively cheap spaceflight routine (if corporate competitors allow them). Large microwave reflectors in orbit could allow for renewable power facilities to be built in remote places — wind farms in Patagonia for example, where wind blows strongly and nearly continuously, but there are few people. They could facilitate energy sharing between geographically distant human communities. The power generated could then be beamed up to the microwave reflector and then beamed back down to an antenna array on Earth, called a "rectenna" (rectifying antenna) near a populated area. The technological "show-stoppers" to this kind of project were resolved during the 1970s and 80s.[298] [299] [300]

---

296. Virgin Galactic, http://www.virgingalactic.com/

297. SpaceDev, http://www.spacedev.com/newsite/templates/homepage.php?pid=2

298. Space.Com, 17 Oct. 2001, Bright Future for Solar Power Satellites, http://www.space.com/businesstechnology/technology/solar_power_sats_011017-1.html.

An even more ambitious power agenda would consist of actually capturing sunlight in space by very large satellites located in geosynchronous orbit.[301] Materials costs could be massively reduced by essentially catapulting raw material for construction from the lunar surface, as was first proposed in the 1970s by the late Princeton University physics professor Gerard K. O'Neil in his book *The High Frontier*.[302]

None of this would change the fact that a small group of corporations has a lock grip on the principal energy sources. In fact, countries that still have some fuel resources of their own but who are not active in space technologies would be completely hostage to the few who are. Who would set the rules, the prices, the quotas?

If such a project could be pulled together, it would take advantage of the full power of between-group cooperation on a global scale. Now, that kind of Coalition of the Willing would be one worth joining.

The values promoted most heavily by the corporate-hijacked American culture, which has been dominant in the world for over half a century, are the root cause of the coming crisis; and those same values are the main obstacle to any real solution. A return to, or a more sincere implementation of, values based upon sustainability and cooperation would much improve humanity's chances of survival.

There are not, and cannot be, purely technological solutions to our imminent crises. The ultimate cause of these crises originates from our civilization's core values. Rationally changing these values is required to find solutions to them.

---

299. *Space Daily.com*, Aug. 11, 2003, The Case for Solar Power Satellites, http://www.spacedaily.com/news/ssp-03b.html

300. Brown, W. C., The History of Power Transmission by Radio Waves, in: *Microwave Theory and Techniques*, Sept. 1984, volume 32, issue 9, PS: 1230-1242, http://ieeexplore.ieee.org/xpl/abs_free.jsp?arNumber=1132833

301. Rutgers.edu, Solar Power via the Moon, Reproduced from American Institute of Physics, *Industrial Physicist*, April /May, 2002., Volume 8, Number 2, PS 12-15, http://www.tipmagazine.com/tip/INPHFA/vol-8/iss-2/p12.pdf

302. O'Neil, Gerard K, *The High Frontier: Human Colonies in Space*, 3rd Edition, Apogee Books, Wheaton, IL, 2000.

PART III
SURVIVAL & REBIRTH

Chapter 11. Seeking Change

Few people are capable of expressing with equanimity opinions which differ from the prejudices of their social environment. Most people are even incapable of forming such opinions. — Albert Einstein[303]

All signs indicate that we are already entering a phase of sharp reductions in fuel availability, not to mention water. Higher prices at the gas pump are just the tip of the iceberg, a form of rationing that at the same time brings profit to the oil companies. Unless a brilliant solution is found and developed and implemented on a grand scale, and fast, we can be fairly sure that we'll need to change our expectations. Many of the things we take for granted may become rare luxuries. The more unrestrained we've been, the more the adjustments will hurt. Start getting used to the idea!

Meanwhile, how can we accomplish political transformation at the scale that is needed? The late Saul Alinsky, a labor union and civil rights activist, spent a lifetime developing just the repertoire of tactics that are needed. In his seminal work *Rules for Radicals*[304] he lays out basic rules for those who would instigate fundamental change. Alinsky observes that revolution is a profound and fundamental transformation of society. It need not be a violent or military process. It can occur in the hearts and minds of citizens. This is the essence of "People Power." Revolution starts from *within* the existing political system.

People must first be made to give up on the existing system before they will become receptive to fundamental change. They must understand that the existing system will not improve their lives and the lives of their children before they will be willing to embrace fundamental change. Alinsky's phi-

---

303. The Quotations Page: Albert Einstein, http://www.quotationspage.com/quote/9049.html
304. Alinsky, Saul, *Rules for Radicals*, Vintage Books, New York, NY, 1971.

losophy is based on optimism: when people realize the need to change, they can accomplish it!

Consider what John Adams candidly said about the genesis of the American Revolution:

> The Revolution was effected before the war commenced. The revolution was in the hearts and minds of the people.... This radical change in the principles, opinions, sentiments and affections of the people was the real Revolution.[305]

Fundamental social change has always been revolutionary — not evolutionary. An existing power relationship is altered, leading to new social relationships. Alinsky states:

> Radicals must be resilient, adaptable to shifting political circumstances and sensitive enough to the process of action and reaction to avoid being trapped by their own tactics and forced to travel a road not of their own choosing. In short radicals must have a degree of control over the flow of events.[306]

To change society, we must understand its broad composition as well as the stake each social class has in maintaining society as it is. According to Alinsky, society is divided into three groups:

1) The haves (the upper-class).
2) The have-a-little-want-mores (the middle class).
3) The have-nots (the underclass).

The "haves" control society politically, economically, and socially. They "create" values which justify their control and delegitimate attempts at fundamental change. The media and social institutions (including schools) indoctrinate the population in these "values," in a subtle form of brainwashing, so that it seems natural and right to maintain the status quo. Alinsky notes that: "All revolutionary ideas are condemned as being immoral, fallacious, and against God, country, and mother."[307]

At the bottom of society, the "have-nots" are disillusioned, demoralized, poorly educated, and focused on survival. They are repressed by a variety of means (the "War on Drugs," police brutality for minor misbehavior, etc.).

The "have-a-little-want-mores" are schizophrenic: they fear change since it may deprive them of the material assets they possess; but they desire change since they want greater wealth than they have been able to acquire under the rules established by the haves. They are envious defenders of the status quo, yet they are never really comfortable with it. Their jobs often exist at the whim of the "haves." They both envy and resent this class. It is from the ranks of the "have-a-little-want-mores" that most revolutions have begun.

---

305. Ibid., Alinsky, p 5, #304.
306. Ibid., Alinsky, pps. 6-7, #304.
307. Ibid., Alinsky, p. 7, #304.

In order to effect social change, people must be made to lose faith in the existing system while believing in themselves and their ability to effect change. They must be given a compelling vision of a better social order. The principles of the US Constitution, which embody the ideals of the Enlightenment, provide this.

> *Give me a place to stand, and a lever, and I can move the earth.*
> *— Archimedes*[308]

HISTORICAL TRAJECTORY OF HUMAN CIVILIZATION

Civilizations are systems for energy production and manipulation into wealth, and distribution of wealth. The earliest civilization was built upon the breakthrough in learning how to concentrate solar energy in the forms of cultivated crops and domesticated animals. This breakthrough allowed for more people to be sustained in patterns which rapidly grew in density and sophistication, becoming first villages, then towns; and finally that foundation of civilization itself, the city.

The organized distribution of energy throughout the resulting dense interaction patterns resulted in the emergence of a political economy, eventually regulated by the police powers of the newly created states. Human political economies were inherently more efficient at converting raw energy to wealth because of the power of between-group cooperation leading to the efficiencies of specialization.

Once these early agrarian political economies had developed, a long-term trajectory for human civilization emerged. The direction was towards greater consumption of energy, which was converted into wealth by means of greater societal complexity. Associative learning with information stored, grouped, and retrieved schematically, as is inherent to humans, allowed for the development of knowledge in the form of refined patterns of schematic aggregation increased slowly, by trial and error, and occasional inspiration, in these early civilizations.

For practical purposes, the earth's resources appeared to be infinite, while human activities only affected natural systems in localized areas. For example, millennia of irrigation coupled with an expansion of eventually unsustainable agricultural practices eventually transformed the lush fields of Mesopotamia, where the first cities arose, into what has been described as "a lunar wilderness." This is how the foundational civilization of Sumeria fell.

Periodic shocks such as these could render previously long-established behavioral schemas suddenly maladaptive. The old ways didn't work anymore. Survival considerations then led to rapid innovation and development of new,

---

308. The Lever Quotations: Archimedes, http://www.math.nyu.edu/~crorres/Archimedes/Lever/LeverQuotes.html

often discontinuous, group behavioral schemas. New knowledge, new values, new cultures were born as a result.

Yet throughout this long period of six thousand years during which agriculture was the energy foundation of civilization, human systems remained decoupled from natural ones. And the belief that humanity could take whatsoever it desired of the fruits of the earth without consequence became axiomatic.

The Industrial Revolution changed all of that within the span of several human lifetimes.

The energy contained within hydrocarbons is immense. It has been estimated that a single barrel of oil "contains the equivalent [energy] of almost 25,000 hours of human labour." Assuming 2,080 hours for a typical work year, that's about the energy equivalent of twelve people working for an entire year, contained in just one barrel of oil.[309] Immense new vistas opened up. For the first time, a global civilization was created, as Early Modern European civilization spread across the planet through imperial conquest. Five centuries on from that outburst, industrial civilization — global capitalism — has overtaken the planet.

The beginnings of agriculture and settled life ten millennia ago led inexorably to the emergence of global civilization. When humans evolved a certain type of knowledge-based, mercantilist, and expansionistic culture and unlocked the energy bonanza stored in the planet's hydrocarbon reserves, all of the planet's other civilizations would either be subsumed, like the Pre-Columbian Americans, or would have to adapt the new cognitive and material innovations to survive, as Japan, China, and India did.

Many trajectories through learning space were still possible at this point. The key variable was the ability of industrial civilization to learn from natural reality and to adapt those lessons to the methodologies of wealth creation. Unfortunately such learning did not occur. National and corporate policy did not take into account the effects of humanity upon the biosphere and upon the planet's natural cycles. And so one particular trajectory was selected, despite its grave shortcomings.

Humanity's wealth-producing, ace-in-the-hole methodology of between-group cooperation unfortunately came to be exploited for the ultimate concentration of wealth in the hands of the few: those who controlled the engines of global wealth creation, the multi-national corporations. Once these artificial entities were declared "persons" at law, mankind's fate was sealed. Human civilization had been hijacked by nonhuman entities that cannot see how their actions affect life on earth. They only measure and can only see competition and profit. The trajectory now leads towards an attractor which represents the mul-

---

309. *Times online: The Sunday Times*, Oct. 16, 2005, Waiting for the Lights to Go Out, http://www.timesonline.co.uk/article/0,,2099-1813695_1,00.html

tiple interlocked crises of resource depletion, political and economic failure, climate change, and ecological devastation and disruption.

To get off this disastrous trajectory and plot a better course, we must overturn the status quo. We have found a place to stand. We have our lever.

PROBLEMS AND SOLUTIONS: TRIUMPH OF MONEY AND POWER OVER DEMOCRACY IN THE US

Last guys don't finish nice. — Saul Alinksy[310]

At the heart of the question of how and why US democracy has been increasingly subverted is the interface between money and the outputs of the political system. While money has always been essential to gaining political office in the US government, there have been throughout American history competing sources of funding which allowed for real if somewhat narrow differences in governmental policy to be placed before the American electorate.

The US electoral system is quite restrictive in that its "winner take all," "first past the post," single district (only one candidate is elected per electoral race for each electoral district) electoral system fairly compels a one or two party system. Thus, smaller parties which would be represented in the government in a parliamentary system are marginalized. Having only two electable political parties means that monetary inputs into the political system are highly concentrated in the hands of a very few players.

While the inherent structure of the US political economy may have provided opportunity for the wealthy to consolidate their position in society and to dominate the political forum, in contradiction to the much-touted precepts of equal opportunity and representation, this takeover did not occur by accident. Rather, it was a reaction at the top against a hundred years of social and political advances for the average family in the US, advances which were gained in part by mass actions of ordinary people struggling to secure better living conditions as the developed countries began to step back from the most onerous exploitation of the early Industrial Revolution. What we have now is, in effect, a counter revolution that was carried out by the aggrieved elites.

Corporate control over the US political system strengthened from the end of the US Civil War through the Great Depression. The Progressive Era represented only a brief, partial setback during this long Gilded Age for the major corporate owners. This regressive trend was broken by the Great Depression, which engendered a systemic crisis of the American political economy. This crisis brought Franklin Roosevelt to the US Presidency in 1933. Roosevelt took advantage of the crisis caused by the Depression to fundamentally reform the

310. Wisdom Quotes: Saul Alinsky, http://www.wisdomquotes.com/002573.html

political economy. Broadly, in order to "save" US capitalism, he grafted elements of socialism onto it: Social Security, for example. The size, role, and scope of government became much greater. Labor unions were fully legalized. All of this activity is known of collectively as the "New Deal."

The Democratic Party became the clear choice of Labor, even as the Republican Party became the party of Business. The American electorate knew whom to vote for if they were pro-labor or pro-business. Overall during these years the power of the citizenry increased, although much of that power was expressed collectively and not individually via the mechanism of labor unions and, through them, the Democratic Party.

World War II and the subsequent Cold War continued the trend towards further consolidation of power in the hands of the central government. This period was marked by the emergence of a permanent war economy — something that had never existed before in the US, and which the nation's founding fathers had explicitly warned against.

In his farewell address to the nation in early 1961, Republican President Dwight Eisenhower warned:

> "Until the latest of our world conflicts, the United States had no armaments industry. American makers of plowshares could, with time and as required, make swords as well. But now we can no longer risk emergency improvisation of national defense; we have been compelled to create a permanent armaments industry of vast proportions. Added to this, three and a half million men and women are directly engaged in the defense establishment. We annually spend on military security more than the net income of all United State corporations. This conjunction of an immense military establishment and a large arms industry is new in the American experience. The total influence — economic, political, even spiritual — is felt in every city, every state house, every office of the Federal government. We recognize the imperative need for this development. Yet we must not fail to comprehend its grave implications. Our toil, resources and livelihood are all involved; so is the very structure of our society. In the councils of government, we must guard against the acquisition of unwarranted influence, whether sought or unsought, by the military-industrial complex. The potential for the disastrous rise of misplaced power exists and will persist. We must never let the weight of this combination endanger our liberties or Democratic processes. We should take nothing for granted — only an alert and knowledgeable citizenry can compel the proper meshing of huge industrial and military machinery of defense with our peaceful methods and goals, so that security and liberty may prosper together."[311]

Vast industrial resources were mobilized by the US government to confront the Cold War threat posed by Soviet Russia. Eisenhower saw the danger and could only counsel vigilance. The Cold War also fostered the emergence of secret government agencies such as the CIA, which often serve to further the

---

311. The Dwight D. Eisenhower Library, Farewell Radio and Television Address to the American People, by President Dwight D. Eisenhower, Jan. 17, 1961, http://www.eisenhower.archives.gov/farewell.htm

domestic and international interests of corporations, rather than those of individual human voters.

As H.G. Wells said in his novel *War of the Worlds*: "intellects vast and cool and unsympathetic, regarded this earth with envious eyes, and slowly and surely drew their plans."[312]

And so it was with the US political economy. The Robber Barons and their descendants, and those who would wish to become robber barons, sought to turn back the provisions of the Progressive Era, and most especially of the New Deal.

During the 1950s through the 1970s African-Americans as well as other ethnic minorities began to have a better chance to exercise their voting rights. American women were allowed into the workplace, in fact they were encouraged or even driven to wish to enter the workplace. The post World War II period, through the middle to late 1970s, was characterized by a doubling of mean income for average Americans. Inequality among Americans lessened noticeably during that period.[313]

Under Lyndon Johnson's presidency, social welfare programs reached their greatest extent. Johnson's "Great Society" programs aimed at the complete eradication of poverty in the US, even as his other governmental policies aimed at mandating a color- and gender-blind society.[314] But Johnson, a prisoner of the Cold War military-industrial-complex, was also escalating the Vietnam War. He declined to raise taxes to pay for it, fearing popular resistance that would throw off his social and societal agenda; instead the war was financed on credit. This was a fatal mistake: it would trigger massive inflation and economic stagnation that went on throughout the 1970s. Johnson was succeeded by Nixon in January of 1969.

During the 1960s, for the first time in history, the majority of the children of the working classes attended some form of college.[315] The creation of a "youth culture" that fostered a sense of entitlement, coupled with the institution of a military draft to supply the rising manpower requirement for the Vietnam War, helped fuel political rebellion among the educated youth.

---

312. Wells, H.G., *The War of the Worlds*, Tor Books, New York, NY, 1886, 1987.

313. See research paper No. 2003/28, United Nations University, World Institute For Development Economics Research, Income Distribution Changes and Their Impact in the Post- World War II Period, March, 2003, http://www.wider.unu.edu/publications/dps/dps2003/dp2003-28.pdf. See also: Center on Budget and Policy Priorities, New I.R.S. Data Show Income Inequality is Again on the Rise, Oct. 17, 2005, http://www.cbpp.org/10-17-05inc.htm. See also: The Library of Economics and Liberty, The Concise Encyclopedia of Economics, Distribution of income, http://www.econlib.org/library/ENC/DistributionofIncome.html.

314. *Encyclopædia Britannica*, 2006, Great Society, Encyclopædia Britannica Premium Service, http://www.britannica.com/eb/article-9000857.

On August 23, 1971, a scant two months prior to his being placed on the US Supreme Court by Richard Nixon, businessman Lewis Powell, then a member of eleven corporate boards, wrote a memorandum to corporate leaders which the US Chamber of Commerce distributed to its national membership under the heading, "*Confidential Memo. Attack on American Free Enterprise System.*" Claiming that the free enterprise system was under pressure from left-inclined political leaders and academics, Powell urged business leaders to take the initiative and assert control over four key sectors of American life:

1) Institutions of higher education, particularly social sciences departments and curriculum including textbook content.

2) The media.

3) The political system and government.

4) The court system.

Businessmen such as brewery-owner Joseph Coors responded by funding the first right-wing think tanks. These included the Heritage Foundation, the Cato Institute, Accuracy in Media, and many others. On-campus groups that were established by right-wing activists and supported with right-wing industrialists' money include the Federalist Society, which will be discussed later in this chapter. In fact, a whole new apparatus of corporatist control grew out of a burst of activism that was ignited by Powell's memo.[316][317]

Also in that eventful year of 1971, President Nixon withdrew the US from the gold standard, effectively devaluing its national currency as measured with respect to gold. Then the oil shocks began in October of 1973. Rapid inflation set into the US economy. The US national budget began to show increasing deficits.

Nixon resigned from office to avoid impeachment in August 1974. The Republican Party was devastated in the 1974 and 1976 Congressional elections. Democrats had used their nearly unfettered powers to investigate the workings of the national security apparatus and to constrain future Presidents' war making powers (War Powers Resolution of 1973) during Nixon's final months in office.[318]

---

315. See U.S. Department of Health and Human Services statistics beginning at: http://aspe.hhs.gov/hsp/97trends/eal-l.htm. Scroll to subsequent indicators by clicking on "next indicator" link until end of document is reached. This is a subsection of: Trends in the Well-Being of America's Children & Youth, 1977 Edition, Office of the Assistant Secretary for Planning and Evaluation, U.S. Department of Health and Human Services, 1997, http://aspe.hhs.gov/hsp/97trends/intro-web.htm. See also: U.S. Census Bureau data on school enrolment at: http://www.census.gov/population/www/socdemo/school.html.

316. Reclaim Democracy.org, the Powell Memo, http://reclaimdemocracy.org/corporate_accountability/powell_memo_lewis.html

317. Media Transparency.org, Aug. 20, 2002, The Powell Manifesto, How a Prominent Lawyer's Attack Memo Changed America, http://www.mediatransparency.org/story.php?storyID=21

The power of the Legislative branch of Government with respect to the Executive branch was restored. Nixon's "imperial presidency" seemed to be a thing of the past. The US Supreme Court had elevated the prestige and power of the judicial branch of government by ruling wisely and judiciously during the Watergate crisis, and in Vietnam-related cases such as the freedom-of-the-press issues arising from *Nixon v. N.Y. Times*. In that case, Nixon had sought to suppress publication of incriminating materials concerning US entry into the Vietnam War. American democracy seemed to be working.

Inequality was at historic lows at this moment in history. Income distribution as measured by indices such as the Gini coefficient (an index of how equitably income is distributed) had attained its historically greatest level of equality.[319] Citizen control of government was substantial. Governmental secrecy was discernibly diminishing.

Shocked, humiliated, beaten, and facing collapse, the Republican Party reinvented itself. Its moderates were discredited. Right-wing conservatives with corporate funding had been organized and increasingly active since Republican presidential candidate Barry Goldwater lost his 1964 campaign. Now, in the course of less than a decade, they took over the Republican Party; they were intent upon rolling back the New Deal and the Progressive era.

During Carter's presidency (1977-81) Republicans were temporarily excluded from the levers of power; Democrats were left in control of the government. During this time, which was characterized by rapidly rising oil prices, Johnson's economic chickens came home to roost: stagnation coincided with high inflation; crime shot up; youth seemed out of control; the US was humiliated abroad by Iranian "students"; and the USSR seemed to be on the march from Angola and Afghanistan to Zimbabwe. Carter's presidency, Democratic party rule, and, in time, all of the 20th century's social, political and economic gains were now on the chopping block.

Reagan's election in November of 1980 brought the end of an era. The now revitalized and reinvented Republicans were fueled by corporate money. In 1976 in the case of *Buckley v. Vallejo*, the US Supreme Court had held that giving of money in effectively limitless amounts ("soft money") was protected under the free speech clause of the 1st Amendment to the US Constitution. Nixon had pioneered the use of covert racism to strip the Democratic Party of its once secure voter base among Southern whites. This voting trend accelerated under Reagan. Democrats were floundering.

---

318. US Info, Text of War Powers Resolution, http://usinfo.state.gov/usa/infousa/laws/majorlaw/warpower.htm.

319. *Encyclopædia Britannica*, 2006, Gini, Corrado, Encyclopædia Britannica Premium Service, http://www.britannica.com/eb/article-9036869. See also: About Business and Finance: Economics: Definition of Gini Coefficient, http://economics.about.com/cs/economicsglossary/g/gini.htm

The Republican Party increasingly became Southern-based. As this occurred it became possible to create a working alliance with Southern evangelical "Christians." Calvinist-derived Pentecostal doctrine was integrated into the Republican agenda. These doctrines glorified wealth as a sign of God's favor, thus legitimating those at the top once again. Business was rapidly deregulated, freeing it from governmental control or oversight (i.e. environmental standards, consumer protection, etc.). Governmental secrecy increased and unofficial conflicts like Iran-Contra and other secret wars were waged unconstitutionally — without the consent of the Congress, even without the knowledge of Congress. Many Democratic leaders joined the corporate deregulation bandwagon at this time.

The Iran-Contra scandal[320] did come to light eventually, but few effective measures were undertaken to redress its abuses and excesses. The balance of power between the Executive and Legislative branches once again favored the Presidency, while Congress was divided along ever more vitriolic partisan lines.

During the twelve years of the Reagan and Bush I administrations, an ideologically far-right national media was established and its hold upon the nation's access to information was consolidated. It continued to grow during the eight years of Clinton's Presidency.

The Fox network of ideologue Rupert Murdoch began spreading its influence across the US and beyond. Reverend Sun Myung-Moon's *Washington Times* newspaper and *Insight* magazine became icons of the new right-wing media. Moon was, and remains, a major financial contributor to far-right Republicans.

Reagan appointees to the FCC repealed the "Fairness Doctrine" that had previously constrained the media from partisan one-sidedness.[321] Laws governing the ownership of media were gradually relaxed, and Republican-friendly corporations began to purchase and consolidate greater and greater chunks of the media, fashioning it into a venue to disseminate right-wing views and attack competing views. Rush Limbaugh was a pioneer of these techniques. Talk radio became a favorite forum for the far right during this period.

Richard Mellon Scaife, a billionaire, following the lead of Joseph Coors and other right-wing businessmen, poured funds into the creation of vociferous attack organs such as his *American Spectator* newspaper. Moderate and especially liberal ideas were attacked; and increasingly, people espousing any principles inconsistent with the far-right agenda (even "moderate" Republicans) were subjected to non-stop, orchestrated, often slanderous, public vilification campaigns.

---

320. *Encyclopædia Britannica*, 2006, Iran-Contra Affair, Encyclopædia Britannica Premium Service, http://www.britannica.com/eb/article-9042741.
321. MBC: The Museum of Broadcast Communications, Fairness Doctrine, US Broadcasting Policy, http://www.museum.tv/archives/etv/F/htmlF/fairnessdoct/fairness-doct.htm.

The American Enterprise Institute and the Heritage Foundation served to incubate ideas, tactics, and strategies required to further the far-right's agenda. They also produced "alumni" who were reliable to staff the thousands of president-appointed positions in the federal bureaucracy. These alumni also began to make up an increasing percentage of Republican Congressional staffers. Those Republican members of Congress who could not adapt were retired and were increasingly replaced by loyal cadres. The Democratic center and Left had nothing to match and counter this.

The Federalist Party, founded long ago by Alexander Hamilton, had desired a strong presidency that functioned like a monarchy. Its ideological descendent, the Federalist Society, was founded in 1982 to identify, train, and nurture a new generation of attorneys who could become the legal arm of corporate America; indeed, they would help to ensure that corporate America's agenda and the right wing's agenda were one and the same. More importantly, the Federalist Society would nurture and indoctrinate suitable candidates for appointments to the federal judiciary.

This effort would culminate in 2005-2006 with the appointments of Federalist Society alumni John Roberts as Chief Justice of the Supreme Court, followed rapidly by the appointment of Samuel Alito as Associate Justice of the Supreme Court. These join Antonin Scalia, who was one of the founders of the Federalist Society; and Clarence Thomas. Associate Justice Anthony Kennedy is considered a close affiliate of the Federalist Society, giving that exclusive group a solid conservative bloc of four Justices on the Supreme Court with a close affiliate, Kennedy, making up a potential deciding fifth vote.

In the case of Hamadan v. Rumsfeld[322], decided on June 29, 2006, Kennedy defected from the conservative bloc, handing the Bush administration a stinging judicial defeat in a case relating to the scope of presidential authority. However, the case also demonstrated the solidity of the four-justice conservative block. One additional Bush appointment to the court will produce a solid conservative majority. As it is now, however, Kennedy only has one foot in the conservative alignment. Kennedy sided with the conservative bloc in the case of Garcetti et al. v. Ceballos,[323] which substantially reduced the First Amendment rights of governmental whistleblowers — furthering governmental secrecy even when the secret activity is unlawful. Kennedy also voted with the four conservatives in the case of Hudson v. Michigan[324], which heavily eroded the Fourth Amendment protections which have been a part of American jurisprudence since the late

322. *MSNBC.com*, Hamadan v. Rumsfeld, Secretary of Defense et al, http://msnbc-media.msn.com/i/msnbc/sections/news/060629_Hamdan_Rumsfeld.pdf

323. *MSNBC.com*, Garcetti v. Ceballos, http://www.supremecourtus.gov/opinions/05pdf/04-473.pdf

324. Supreme Court of the United States, syllabus of Hudson v. Michigan, http://www.supremecourtus.gov/opinions/05pdf/04-1360.pdf

1700s. Kennedy is very much the "swing" vote on the Supreme Court. It is upon the slender thread of Justice Kennedy's *sometimes* adherence to long-established interpretations of the US Constitution that what remains of the rule of law continues, for now, to persist in the US.

There are three levels in the US federal court system: the District Courts, the Appellate Courts, and the US Supreme Court. According to Article III of the US Constitution, federal judges serve for life. They are nominated by the President and are then subject to a vote by the Senate. A simple majority is required for approval.

The strategy of the far right was, and is, to rush as many "ideologically reliable" judicial nominations as possible through the Senate when they can. Any and all means are utilized to accomplish this court-packing goal. For example, right-wing propaganda organs (increasingly this means the *entire* corporate-consolidated US media) lambaste and vilify all senators who oppose a Republican nominee, while repeatedly depicting that nominee in glowing tones in the media. Should a Democrat occupy the White House (Clinton), the strategy shifts to one of limiting the number of Court appointments that are made. This creates numerous vacancies which can be subsequently exploited by the Republicans when they re-take the Presidency and claim that there is a "crisis" in the judiciary.

As of 2006, Republican appointees control about seven of the US's thirteen Courts of Appeal; Democrat appointees control possibly four; control is divided for the other two. This balance shifts relentlessly towards the far-right with each new Federalist Society approved judicial appointment.

A vast network of religious groups which actively — and unconstitutionally — indoctrinate and instruct their tens of millions of worshippers to vote for right-wing Republicans, acts in concert with corporate-owned religious television and radio networks.

Economically, the US has not seriously attempted to enforce the provisions of the Sherman Anti-Trust Act of 1890 since the time of Reagan's presidency. The US market is oligopolistic: a few big firms control each market sector and effectively prevent real competition. These firms are characterized by interlocking Boards of Directors. They funnel vast amounts of corporate monies into the political process, ensuring the election of "business-friendly" candidates.

Some degree of split has emerged between "old" industry (mining, energy, construction, etc.) and "new" industry (computers, software, pharmaceuticals, etc.) market sectors. The old industry market sectors are firmly onboard the far-right Republican juggernaut. Within the new industry camp, there is a clear perception that their corporate interests and those of the old industries diverge. Some of these new industry elites have begun to move to capture the Democratic Party, in order to fashion it into a vehicle to further their divergent (from the old industry elites) interests. This intra-elite split may offer a glimmer of hope, if it can be exploited. Corporatist supporters in the Democratic Party often refer to themselves as "New Democrats." They represent both old industry and the new industry factions. Their coexistence in the Democratic Party is thus always uneasy.

Coordinating all of these divergent interests are shadowy individuals such as Grover Norquist. The Washington D.C.-based Norquist appears to act as a coordinator for the various right-wing entities. In his recent book, *Blinded by the Right*, David Brock, a defector from the right wing, describes Norquist as regularly holding court at the head of a long table in the salon of his D.C. home, where a large portrait of V.I. Lenin was displayed! According to Mr. Brock, Norquist is attracted by Lenin's writings on the seizure of absolute power by a small core group, and Norquist boasts publicly to his guests about his wish to emulate his famous "mentor."[325] [326]

During his eight years as president, William Jefferson Clinton, a "New Democrat" corporatist himself, was pinned down by merciless and unrelenting fire from the right. His presidency, though not without accomplishment, was forced into a more or less permanent defensive holding action by the right-wing juggernaut.

George W. Bush's selection as president in 2000 by a clearly partisan (and unconstitutional) 5-4 majority decision of the US Supreme Court, along with the legally questionable recapture of the Senate by the Republicans in the 2002 elections and Bush's equally questionable reelection, now leave the far-right in effective control of all levers of power in the US.[327] [328] The Democratic opposition has been under such strenuous attack for years that most Democratic senators and Congress members will not speak out against Republican outrages.

The media have become ever more consolidated and ever more favorable to the far-right.[329] Under the former Chairmanship of Michael Powell (son of Colin Powell), this trend continued apace, as was clearly shown by the FCC's 3-2 vote to further augment the process of media consolidation on June 2, 2003.

Paul Krugman, *New York Times* columnist and Princeton University economist, notes that the US Treasury is being *willfully* bankrupted to destroy any possibility of this country maintaining any social welfare programs anytime in

---

325. *The Nation*, May 14, 2001, Grover Norquist: "Field Marshal" of the Bush Plan, http://www.thenation.com/doc/20010514/dreyfuss

326. *Votelaw.com*, How Much is a List Worth? http://www.votelaw.com/blog/archives/001153.html

327. *In These Times*, Feb. 15, 2005, A Corrupted Election, Despite What You May Have Heard, the Exit Polls Were Right, http://www.inthesetimes.com/site/main/article/1970/

328. *Election Archive.org*, Investigating the Accuracy of Elections, http://www.uscountvotes.org

329. *PBS.org*: NOW, Transcript: Bill Moyers Interviews Barry Diller, http://www.pbs.org/now/transcript/transcript_diller.html

the 21st century.[330] There will be no Social Security, no Medicare for the aged, and no adequate veterans' benefits.

Nationwide, mass religious institutions now serve to legitimate the US's ruling party. This is reinforced by ever more consolidated media, which are ever more firmly in the ideological hands of the far-right.

The US military is increasingly politically homogenous, particularly at the officer levels. Defense Secretary Rumsfeld now personally approves or vetoes all promotions to flag rank (General or Admiral), in part on the basis of ideological reliability.[331] [332]

Within the US federal government, the system of institutionalized checks and balances is breaking down. The federal judiciary increasingly exists to "legitimate" the actions of the ruling party. Congress is a rubber stamp for the Executive branch. The President is increasingly above the law and religion is used to effectively anoint the president. Money and power now form a nearly integral whole.[333] The "divine right of kings" is returning. Along with all this the terrorist attacks of September 11, 2001 have been used to create a climate of fear and militant "patriotism."

The social and economic reforms of the 20th century are about to be lost, along with democracy itself. We are on the edge of an abyss, sliding toward a nightmare of corporatism, fascism, runaway global warming, and resource depletion.

If there is any chink in the corporate elite's armor, it is that split referred to above, a split between old industry and new industry. The old industry market sectors are firmly onboard the far-right Republican juggernaut. New industry's corporate interests are somewhat different, and their political inclination leads them to the Democratic Party, which they could logically use as a vehicle by which to counter Old Industry.

Old Industry, the chemical and pharmaceutical industry, energy, mining, and construction, aerospace and military sectors, are solidly behind the Bush Administration. Defense contractor General Electric owns NBC, CNBC, half of MSNBC, and more. Other sections of the information sector are also coming rapidly under the control of these sectors. Inevitably, the facts of corporate ownership must color the news —what is presented as well as how it is presented — and other programming decisions.

---

330. *New York Times*, April 29, 2003, Matter of Emphasis, http://www.nytimes.com/2003/04/29/opinion/29KRUG.html?ex=1136955600&en=6a063283a0abe9c1&ei=5070

331. *Military.com*, April 26, 2003, Joe Galloway: Army Shakeups Clear Path for Rumsfeld's Vision, http://www.military.com/NewContent/0,13190,Galloway_042603,00.html

332. *Daily KOS*, Dec. 29, 2005, Bush Changes, Politicizes Pentagon Succession, http://www.dailykos.com/story/2005/12/29/2522/2369

333. *In These Times*, Feb. 18, 2005, The People's Business: Controlling Corporations and Restoring Democracy, http://www.inthesetimes.com/site/main/article/1971/

Industries that backed Bill Clinton include bio-tech, electronics, computers, software, financial services, and the like.[334] Clinton was a "New Democrat" and a corporatist, however; it's just that he represented the new-industry faction of corporatism. He and G.W. Bush both are supporters of corporatism — they could not have gotten elected otherwise. Still, the policy differences between them are much more than cosmetic.

Both the "old" and the "new" industries manifest themselves as politically influential corporations. Both are beholden to the present political economy in which vast concentrations of money and human resources (corporations) control the outputs of the political system. Stated another way, both are equally committed to the disenfranchisement of citizens, flesh-and-blood people, to the benefit of corporations.

The Republican corporate oligarchs are almost entirely linked to highly polluting industries that may already be past their hey day. Their products are prized chiefly for the value of the matter, as well as unskilled or low-skilled, physical labor that they contain. New industries are knowledge intensive. They use a relatively small amount of raw material to produce products which are valued almost entirely for the knowledge which was used to transform them into a useful product.

All of them have an interest in maintaining the present system of political economy, based as it is upon corporate rule. However, old industry Republican oligarchs fear the inevitable sunset for the age of oil, and wish to prolong it no matter what the costs to humanity and the ecosystem. New knowledge-intensive technologies threaten their position of dominance and they ultimately have different fundamental interests. To maintain their hold, the Republican right increasingly uses the media to promote their one-sided view of things and they also promote a "Dominionist" faux Christianity that celebrates subordination of individuals to self-proclaimed "Godly" leaders and embraces environmental devastation as "proof" that their Messiah will return soon. Since most of the earth's petroleum reserves are in or near the Middle-East, it is easy to foment racial and religious hatred to sell what is actually an imperial energy war to the population.

New technology industries depend upon educated scientific researchers and ambitious entrepreneurs who are unconstrained by ideology or by social and societal convention. For this grouping, religious dogma is an impediment to many new industries — particularly in the biotechnology sector. Indeed the entire old technology sector is an impediment to their interests.

Old technology industries place much less emphasis upon fundamental innovation and much more emphasis upon protecting their market sectors. This brings them into potential conflict with those who produce better techniques

---

334. *The American Prospect Online*, April 23, 2001, How The DLC Does it, Volume 12, Issue7, http://www.prospect.org/print/V12/7/dreyfuss-r.html

and technologies to do what the old industries do — and who are seeking to capture market share from the old industries. Old industry workers are supposed to conform to social and societal convention. They are expected to know, and remain in, their place in the corporate, social, and political hierarchies.

It's unsurprising that the Bush administration has provided tax breaks for old industries but has not assured a supply of investment capital to the new industries. For the old industries, there is no tomorrow. Tomorrow belongs to the new industries, but there can be no tomorrow unless the old industries are dislodged.

As America became increasingly ruled by corporate elites, a split began to emerge between the more Libertarian and "New Democrat" leaning new-industry elites, and the fear-mongering old-industry elites. Old industry captured the Republican Party from traditional conservatives quite thoroughly by the 1980s. Since then, they have moved aggressively to concentrate the information industry into ever fewer hands, and then to purchase full control over those hands. This has turned the information industry into an instrument of old industry.

The Christian Rightists were nurtured and indoctrinated to produce a compliant, submissive, and obedient mass voter base for the corporatists. With a large, reliable base, fear and greed — and, if necessary, vote fraud — can be utilized to ensure unbreakable political control over the US, even while the facade of electoral democracy is maintained.

Even the illusion of a two party system can be maintained by cultivating compliant leaders within the Democratic Party. This is the task of the Democratic Leadership Council (DLC) — to ensure effective corporatist control over the Democratic Party. During the author's 2004 campaign for US Congress, Democratic Party Chair Howard Dean told him, "The DLC is the Republican wing of the Democratic Party."

With G.W. Bush's ascendancy, the new-industry elites were forced to defend themselves. They did this by redoubling their efforts to gain control over the Democratic Party. While still supporting corporate control, they require an educated, innovative, and creative workforce. Also, new-industry elites are culturally products of the Enlightenment. They tend to believe in the concept of the self-created autonomous individual.

A vigorous if fairly small political party whose ideology overlaps somewhat with the new-industry Democrats is the Libertarian Party. Libertarians espouse the free market ideology and seek to promote human freedom both socially and politically, as well as economically, without government interference (or protection for the less successful). Libertarian idealism thrives in the new technology sectors. On the new technology frontier, entrepreneurial creativity combined with venture capital, hard work and good luck allow for something like a Smithian Free Market to actually exist. Under these conditions, the tenets of Libertarianism actually approximate economic reality. However, this cutting-edge economic frontier is only a small portion of the economy, and like all fron-tiers it becomes settled in time. In this metaphorical "frontier" context, "settled" means that it becomes oligopolistic, just like the rest of the economy. So Liber-tarianism is a niche political philosophy favored by certain entrepreneurial pio-neers; but it is not, and cannot be, a pragmatic political philosophy for organizing the overall American political economy. The Libertarian Party repre-sents one political interface between this entrepreneurial class and the overall political economy.

Thus, it is the fundamental split between the old-industry and new-industry corporate elites that offers the single most potent opportunity for intro-ducing change in the political system during the coming years. And profound change must be introduced if civilization is to head off the coming debacle.

WHAT IS TO BE DONE?

The present corporate-led political arrangement in the broadest sense rep-resents a backlash against the deepest ideals of the Enlightenment, and of modernity itself. People's fundamental and "inalienable" rights are indeed being alienated, and the people themselves are considered to be fungible goods, lacking in dignity or constitutional rights: consumers and employees to be controlled, used, and discarded at a whim.

The current worldview is intolerant, hierarchical, and hearkens back to feudalism. It is a worldview in which the few rule without question over the many, on the basis of hereditary authority as legitimated by an ideologically compatible right-wing religious establishment which represents the sole arbiter of truth.

At its core, however, this worldview is based upon fear of loss, and clinging to the past. What can be done to turn to a worldview that enables us to embrace change and build a future?

Democrats now represent an uneasy coalition of "old" and "new" Demo-crats. The old or traditional Democratic core is a progressive Enlightenment-descended political grouping in the tradition of Franklin Roosevelt. These people favor citizen rule and a significant public sector for the purpose of pro-viding as nearly equal as possible initial opportunities for individuals to flourish. Entrepreneurism, in the context of these equalizing public goods, is encouraged. These Democrats also increasingly understand the need to operate the political economy in an ecologically rational manner.

The "new" Democrats represent an important interface between corpo-ratists — new and old — and the Democratic Party. Many — certainly not all, however — of these "New Democrats" strongly reflect the interests and values of the new-industry elites. Their values, unlike those of the old-industry elites, are forward-looking and are compatible with many of the Enlightenment-derived values of modernity. There is support for a public sector, though not to the same degree as is the case with the traditional Franklin Roosevelt-type citizen- and worker-centered "Old Democrats."

Because the new-industry sector is heavily populated by successful entrepreneurs and executives of entrepreneurial firms, a belief in the core Enlightenment value of the self-creating autonomous individual is quite strong. Old industry influence exists here as well, of course. However, it is relatively weaker than in the Republican Party. "New" Democrats see the Republican Party generally as something to be blocked and undercut, while they might be able to collaborate with Old Democrats because the old industry elites do not control them.

One could hope to engage the new industry-oriented New Democrats in fighting for the changes that would lead to a more responsible future, but there is a major difficulty. This group is fearful that any weakening of the system could negatively affect its interests vis-à-vis the old-industry elites. Thus, they remain staunch defenders of corporatism.

Green Party activists are the most zealously committed reformers. They seek to shift the political economy away from a one-dollar/one-vote system to one based upon one-person/one-vote; and also to refashion the political economy in an ecologically rational manner towards sustainability. Greens generally believe in an even larger public sector than do the old Democrats.

The Libertarians represent a separate interface between the new industrial sector and the political economy. Overall, they are more zealous members of the same entrepreneurial class as the new Democrats; they tend to reside somewhat further out along the economic frontier. This commonality between the two groups should be noted. These are the people who most strongly see the economic potential of an entirely new "green" economic frontier, comprised of new renewable energy technology ventures.

Libertarians believe in a small public sector, as they believe that markets are more efficient than are governments in allocating resources. Accordingly, Libertarians are well suited for roles involving entrepreneurialism — such as starting companies to produce renewable energy technologies.

One hope for the future is that a counter-coalition can be built to rein in the corporations enough to allow for an alternative political economy to be constructed, based upon different memes. Divisions within the corporatist opposition must be identified and exploited. Forward looking members of the opposition must be persuaded to shift allegiances. Much as war has served as a driver of rapid technological innovation, so a broad recognition of the coming crisis may drive rapid innovation while fostering group cohesion. Also, and equally important, there must be rapid social innovation as well. After all, it was our non-adaptive manner of thinking which precipitated the crises. And of course, it is vitally essential to ensure that the legal doctrine of corporate "personhood" will have died during the transitional period.

If a progressive coalition of Greens and "old" Democrats can ensure the nomination of progressive candidates, then forward-thinking new-industry "new" Democrats, along with libertarians, political independents and rationally inclined fiscally conservative Republicans, can and should be brought into a political coalition created for this purpose. With the money — and the power — of these groups combined, and their various skills, it may be possible to change civilization's course.

***

This work began with the assertion that we live at the decisive moment in all of human history — decisive not only for one culture or another, not only for the "developed world" but for all of humanity. This may have appeared to be an extravagant claim, but the information in this book shows it to be a factual one. Humanity is faced with immense challenges. However, the chance to create a long-term ecologically sustainable and humane future also represents a correspondingly immense opportunity.

In the coming years we must all act and think flexibly and creatively if these challenges are to be surmounted. If we can rise to this occasion we will bequeath an open-ended future to civilization and to the biosphere. This book has focused upon analysis and description of the inter-linked crises facing our civilization. In a subsequent work, I will focus upon prescription — specifically upon strategies for dealing effectively with these crises and thereby creating a positive future.

The promise of the future is what, consistent with the ancient legend of a new world following a global deluge, I refer to as "Infinity's Rainbow."

I'd like to end this work where I began it, with this quote:

> It has often been said that, if the human species fails to make a go of it here on the earth, some other species will take over the running. In the sense of developing intelligence this is not correct. We have or soon will have, exhausted the necessary physical prerequisites so far as this planet is concerned. With coal gone, oil gone, high-grade metallic ores gone, no species however competent can make the long climb from primitive conditions to high-level technology. This is a one-shot affair. If we fail, this planetary system fails so far as intelligence is concerned. The same will be true of other planetary systems. On each of them there will be one chance and one chance only.[335]

This is our one and only chance. Let's not blow it!

---

335. Ibid., *Of Men and Galaxies*, Prologue, #1.

Epilogue

Ah, Love! could thou and I with Fate conspire
To grasp this sorry Scheme of Things entire!
Would not we shatter it to bits — and then
Re-mould it nearer to the Heart's Desire!
— Omar Khayyam[336]

Two apparently unrelated changes (each one improbable enough on its own) set the trajectory of human civilization in its present disastrous direction: we discovered how to exploit the energy locked up in hydrocarbon deposits; and the legal enfranchisement of corporate personhood led to ordinary humans losing control over their political economy.

These corporate entities represent a type of complex adaptive system, which organizes the individual abilities of numerous humans towards a collective goal. Since these entities exist only to produce short-term profits for their investors, that goal has been the creation of wealth as quickly as possible, as cheaply as possible. This means that externalities such as the effects of wealth extraction upon the earth's biosphere are disregarded to the maximum extent legally possible.

Since governmental regulation is the only check upon what is legal for these corporate entities, it follows that corporate profit maximization requires the subordination of this other human system — government. Human laws that stood in the way of corporate profit maximization have been subverted, as corporations gained power over governments.

In pursuit of corporate goals, it looks as though we have far exceeded the planet's natural carrying capacity. Only hydrocarbon energy allows, for the moment, for humanity's billions to continue multiplying. And if the supply of hydrocarbon energy is about to enter into permanent decline, the ramifications are beyond prediction.

---

336. *Rubiayat of Omar Kayyam*, Fitzgerald translation, http://www.okonlife.com/poems/page3.htm

READING LIST

Readers wishing to learn more about the topics covered in this book should consider reading the following books:

I) Peak Oil/hydrocarbon depletion & consequences.

Simmons, Mathew, *Twilight in the Desert, The Coming Saudi Oil Shock and the World Economy*, John Wiley & Sons, Hoboken NJ, 2005.

Kunstler, James, Howard, *The Long Emergency, Surviving the Converging Catastrophes of the Twenty-First Century*, Atlantic monthly press, New York, NY, 2005.

Goodstein, David, *Out of Gas, The End of the Age of Oil*, W.W. Norton & Company Inc., New York, NY, 2004.

Deffeys, Kenneth S., *Beyond Oil, The View From Hubbert's Peak*, Hill & Wang, A Division of Farrar, Straus, & Giroux, New York, NY, 2005.

Heinberg, Richard, *The Party's Over, Oil, War, and the Fate of Industrial Societies*, New Society Publishers, Gabriola, Island, BC, Canada, 2003.

Savinar, Matt, *The Oil Age is Over, What to Expect as the World Runs out of Cheap Oil, 2005-2050*, Morris Publishing, Kearney, NE, 2004.

The Final Energy Crisis, McKillop, Andrew & Newman, Sheila, Eds., Pluto Press, Ann Arbor, MI, 2005.

Klare, Michael T., *Blood and Oil, the Dangers and Consequences of America's Growing Dependency on Imported Petroleum*, Metropolitan Books, Henry Holt and Company, New York, NY, 2004.

Darley, Julian, *High Noon for Natural Gas, The New Energy Crisis*, Chelsea Green Publishing Company, White River Junction, VT, 2004.

Hartman, Thom, *The Last Hours of Ancient Sunlight, The Fate of The World And What We Can Do Before It's Too Late*, Three Rivers Press, New York, NY, 2000.

Heinberg, Richard, *Powerdown, Options and Actions For a Post-Carbon World*, New Society Publishers, Gabriola, Island, BC, Canada, 2004.

Roberts, Paul, *The End of Oil, On the Edge of a Perilous New World*, Houghton Mifflin Company, New York, NY, 2004.

Clark, William R., *Petrodollar Warfare*, New Society Publishers, Gabriola, Island, BC, Canada, 2005.

II) Civilizational Collapse

Diamond, Jared, *Collapse, How Societies Choose to Fail or Succeed*, Viking Penguin, a Member of the Penguin Group (USA) Inc., 2005.

Tainter, Joseph, *The Collapse of Complex Societies*, Cambridge University Press, Cambridge, United Kingdom, 1988.

Wright, Ronald, *A Short History of Progress*, Carroll & Graf Publishers, New York, NY, 2005.

Greer, John Michael, *How Civilizations Fall: A Theory of Catabolic Collapse*, Unpublished manuscript, 2005, On web at: http://media.anthropik.com/pdf/greer2005.pdf#search=%22How%20civilizations%20fall%22, also at: http://www.xs4all.nl/~wtv/powerdown/greer.htm.

III) Systems Theory

Jervis, Robert, *System Effects, Complexity in Political and Social Life*, Princeton University Press, Princeton, NJ, 1997.

IV) Religion and Politics

Goldberg, Michelle, *Kingdom Coming, The Rise of Christian Nationalism*, W.W. Norton & Company, Inc, New York, NY, 2006.

Phillips, Kevin, American Theocracy, *The Perils and Politics of Radical Religion, Oil, and Borrowed Money in the 21st Century*, Penguin Group (USA) Inc, New York, NY, 2006.

V) Money, Corporatism, and Politics

Hartman, Thom, *Unequal Protection, The Rise of Corporate Dominance and the Theft of Human Rights*, St. Martin's Press, New York, NY, 2002.

Sirota, David, *Hostile Takeover, How Big Money & Corruption Conquered Our Government – and How We Take it Back*, Crown Publishers, New York, NY, 2006.

VI) 9/11

The 9/11 Commission Report, Final Report of the National Commission on Terrorist Attacks Upon the United States, W.W. Norton & Company, New York, NY, 2004.

Griffin, David Ray, *The 9/11 Commission Report, Omissions and Distortions*, Olive Branch Press, Northampton, MA, 2005.

Ruppert, Michael C., *Crossing the Rubicon, The Decline of the American Empire at the End of the Age of Oil*, New Society Publishers, Gabriola, Island, BC, Canada, 2004.

VII) Climate Change

Cox, John D., *Climate Crash, Abrupt Climate Change and What it Means for our Future*, Joseph Henry Press, Washington, D.C., 2005.

Lovelock, James, *The Revenge of Gaia*, Allen Lane, an Imprint of Penguin Books, London, United Kingdom, 2006.

Appendix

Introduction

This article explains the author's concept of crisis-driven evolutionary learning in detail. Many of the key concepts that are discussed in the present book, including crisis-driven learning, reciprocity, and comparative advantage, are dealt with in the context of computer simulation.

It was originally published by Kluwer Academic/Plenum Publishers as Chapter 16, "Global Modeling and International Performance," in The Performance of Social Systems: Perspectives and Problems, edited by Prof. Francisco Parra-Luna, 2000.

In this research the author was seeking to gain understanding of the process of societal learning occurring at a time of rapidly increasing technological ability to inflict ever more lethal harm upon eve more of the world by ever fewer people. It does not specifically investigate the several crises discussed in the previous chapters. By operationalizing key assumptions (essentially by translating those assumptions into BASIC language commands and simple algebraic equations) the author was able to produce a model, without "tweaking," which was able to fairly accurately "retrodict"—that is, predict retrospectively—patterns of international warfare during the past 500 years of the international system.

The mean result of a large number of runs produced about a 50-50 chance that human civilization would be able to learn its way out of using war as a means of attaining state goals before war led to the collapse of civilization. This is, in effect, a study about learning versus political failure. One unsettling consequence of this research is that it identifies a fairly common failure mode for global civilization that, once it got underway, invariably led to total collapse. What happened is that one power would come to dominate the entire model international system. It would have high values for the study's indicator variable for democracy, and call it "fair play" towards others. Then it would "go bad."

Apparently, unconstrained power usually proved to be massively corrupting. This process of going bad would set a series of events into motion in response. The result was always the same: Massive warfare leading to total civilizational collapse planet-wide. This is extremely unsettling because it seems to closely resemble what has actually occurred with respect to the USA in the wake of the Cold War. Apparently power always must be counterbalanced...

CRISIS-DRIVEN EVOLUTIONARY LEARNING AND THE CHARACTERISTICS OF THE INTERNATIONAL SYSTEM

The nature of the international system is investigated. It is found to comprise a complex adaptive "*learning*" system. Pursuant to this inquiry the issue of whether this learning system is best characterized by between-group conflict only, or by a temporally varying combination of between-group conflict and between-group cooperation, is investigated. To accomplish this aim, potential causal relationship(s) between inter-state trade and inter-state warfare, in the context of historically increasing levels of systemic power are investigated. The modern world system is hypothesized to evince crisis-driven learning behaviors with respect to time.

To test these propositions, their underlying theoretical basis is operationalized into a 10-nation BASIC language computer simulation of the world system. Two classes of model systems, one premised upon realist power-dominated systemic assumptions, along with another premised upon the hypothesis that international trade plays a key role in determining the trajectory of the world system are evaluated in this context. Key model variables are adjusted to allow for methodical testing of hypothesized economic interaction (interstate trade) effects upon warfare frequencies of a statistically significant (226) sample N of model system runs. Empirical data for real world warfare frequencies (Levy, 1983) are utilized to allow for potential falsification of model derived findings. Clear and definitive conclusions are drawn from this research.

Key Words

- systemic learning
- complex adaptive systems
- computer simulation
- Ricardo's Law of Comparative Advantage

SECTION I: CONCEPTUAL AND EMPIRICAL FOUNDATIONS OF THE MODEL

The modern international system began to coalesce around five centuries ago. Technological, demographic, political, economic, and organizational factors account for this occurrence (Levy; 1983; Chirot, 1985; Modelski, 1987). Its constituent entities interact via the mechanisms of both trade and warfare. The research questions, which I will investigate for this paper, are: 1) Whether the international system constitutes a complex adaptive, "learning" system. I will utilize real world warfare frequency data (Levy, 1983) to evaluate this proposition. 2) Whether a realist/neo-realist type system or a liberal/neo-liberal/idealist type system most nearly corresponds to the actual characteristics of the international system. I will utilize variants of my computer model in which either military power alone determines the outcome of interactions between nations (realism-type paradigm), or, military power and trade determine outcomes for international interactions (liberalism-type paradigm). These two variants in essence represent between-group conflict only, versus a combination of between group-conflict and between-group cooperation. In doing this I am explicitly positing that trade and war are inversely related. There is broad prior theoretical and empirical support for this proposition. For example, Mansfield (1994, pp. 31) notes: "...moreover there is considerable evidence of an inverse relationship between trade and commerce." I should note that he interprets these findings from a neo-realist perspective, which is not wholly consistent with that which I shall develop below.

In conducting this inquiry, I will articulate and evaluate a model of the international system, which learns with respect to time. I hypothesize that systemic learning, in the context of declining frequencies of interstate warfare, has occurred in response too increasingly rapid gains in total systemic power. These power increases are hypothesized to have caused the adverse consequences of warfare, in general, and great-power warfare, in particular, to evince a corresponding increase. This creates 'selection pressure' upon the system to 'learn' to avoid this ever-increasing danger to its well being, and indeed, to its existence. Systems, which are capable of evincing such adaptive learning, are referred to as complex adaptive systems.

With respect to international trade, I predicate this inquiry upon the explicit assumption that, unlike warfare, which is at best zero-sum, trade is non-zero sum. This is to say that, while five minus three is two in the case of warfare; two plus three can, in effect, be six in the case of trade. This occurs because of what is commonly known as Ricardo's Law of Comparative Advantage. Stated as generally as possible, it asserts that given two (or more) groups each of which produces various goods and/or services with differing degrees of efficiency (cost), trade between the groups is always to the mutual advantage of each group, even if one group produces all of the goods and services which another does, and does so with greater efficiency. As one researcher observes:

> ...its [trade] invention represents one of the very few moments in evolution when Homo sapiens stumbled upon some competitive ecological advantage over other species that was truly unique. There simply is no other animal that exploits the law of comparative advantage between-groups. Within groups, as we have seen, the ants, the mole rats, the Huia birds beautifully exploit the division of labor. But not between groups. David Ricardo explained a trick that our ancestors had invented many, many, years before. The Law of Comparative Advantage is one of the ecological aces that our species holds. (Ridley, 1996, page 210).

Thus, while between-group conflict is an ancient resource allocation mechanism for all living organisms, between-group cooperation, as exemplified by trade, is a unique discovery of humanity. As we shall see, this leads to some interesting predicted consequences for the global system.

The Crisis-Driven Evolutionary Learning Model

I predicate my model of crisis-driven learning upon the hypothesis that fundamental systemic level learning occurs principally in response to crisis. Crisis, in this context is defined as any situation in which the application of previously encoded knowledge fails to resolve some situation in a manner in which harm, or the potential for harm, is averted (Byron, 1996, 1997). I assume that the probability of learning is proportionate to the perceived probability of harm arising from not resolving the situation. For the purposes of this study, which is predicated upon an analysis of relative warfare frequencies, this probability is assumed to be unity.

Crisis-driven learning occurs at the systemic level when states are unable to resolve some problem as a result of their individual, state-level and below, efforts. This provides them with strong incentives to attempt to cooperate with other states to resolve the common problem that threatens them with increasing "harm" (loss of power in the context of the model). If successful, these efforts may lead to the creation of new behavioral norms, and/or international regimes. If unsuccessful, war may result.

I have published specific details regarding this systemic learning model elsewhere (Byron, 1996, 1997). Very succinctly, I theorize that systemic learning begins in the minds of individual decision-makers. When some problem, or difficulty, arises which the application of pre-existent inter-state adjudicative, or behavioral algorithms such as international regimes, consultative bodies (e.g. the G-8), or generalized international norms, are unable to resolve, the system experiences a *crisis*. If the direct, or anticipated, consequences of inaction exceed those of implementing change, then inter-decision-maker agreement is sought. If consensus can be attained, then new behavioral algorithms are incorporated into the structure of the international system, as noted above, in the form of new norms, consultative bodies, or regimes. If this process fails, then conflict between states becomes more likely, as attempts are made to force new behaviors which are most beneficial to a given state's interests upon other, possibly unwilling, members of the international system.

Broadly speaking, the reality of systemic learning is increasingly accepted. For example, Modelski (1994, pp. 324) states, "Successful learning produces structural change." What is at most at issue is the nature, and significance of such learning.

Levy (1994) in his magisterial survey of learning and foreign policy distinguishes sharply between individual level learning and collective/systemic learning. He defines individual level learning as comprising: "a change in beliefs at the individual cognitive level..." Levy, 1994, pp. 278). Regarding group/systemic level learning he states:

> ...the reification of learning at the collective level is not analytically viable. Organizations do not literally learn in the same sense that individuals do. They only learn through individuals who serve in those organizations, by encoding individually learned inferences from experience into organizational routines. (Levy, 1994, page 287).

These definitions appear accurate enough. Where they are inconsistent with my crisis-driven evolutionary learning model is in failing to recognize that individual and group level learning, in the context of the international system, are inseparably bound together. They are aspects of a seamless web of interactions. They collectively constitute different stages of one, holistic, process. Restated, they constitute a unitary complex adaptive system. Furthermore, the difficulties inherent in incorporation of new system level learning, which Levy (1994) articulates, explain why systemic learning tends strongly to be crisis-driven.

The international system thus, can be viewed holistically as comprising a structured linkage between many individual human minds, which is temporally extended. These minds do not exist in isolation, but rather exist in the context of a plethora of social structures, decision-making. Hence, learning for these minds, is substantially influenced by the context in which it occurs. This has two important effects: First, decisions are often less than optimal as the ability to interpret information is circumscribed. Thus learning may be both adaptive, as well as maladaptive. A more precise, objective, definition of the terms adaptive and maladaptive in the sense of relative *fitness* will be utilized subsequently in this study. Second, as individual level learning must occur only in this context, all prior individual learning, which has subsequently been incorporated into the structure of the system affects all current learning. Information flows throughout the international system are characterized both by feedback, as well as feedforward, channels. Given this reality it is reasonable to consider that all learning, at any possible level of analysis, takes place within a unified "design space" as articulated recently by the noted philosopher of science Daniel Dennet (Dennet, 1995). As will be discussed below, a system so constituted, is defined as comprising a complex adaptive system. This is the approach that I shall utilize.

It's the explanatory viability of this model, which shall be empirically tested below.

I should also note that this model of learning is NOT equivalent to a Bayesian learning process. Bayesian updating is a learning model that is predicated upon probability updating. It occurs when prior behavioral probabilities are updated on the basis of observations of phenomenological reality. It *assumes* that the updating process occurs with perfect efficiency. That is, information, though possibly subject to limitations in an entities ability to gather it, is subsequently, at least, incorporated into a revised probability update without distortion, and with optimal efficiency. As I've just described my crisis-driven learning model above, such efficient incorporation of information is essentially impossible as the very structure of the social environment (most particularly culture, e.g. Eckstein 1997) in which such updating must occur virtually ensures that perfectly efficient updating will not occur. Of course, this means that I am not articulating a rational actor predicated model.

Regarding warfare, which occurs with increased probability, in the context of my model, when attempts at decision-maker agreement fail; a given state's propensity towards engaging in warfare is empirically related to its level of democratization (Doyle, 1983, 1986). I hypothesize that the underlying variable is the state's degree of inter-state reciprocity (cf. Axelrod, 1984). Reciprocity correlates with democracy because individual actors possess *innate* goals, including, as Jefferson articulated, "life, liberty, and the pursuit of happiness..." States, which are constituted internally so as to maximize these goals, necessarily possess high degrees of individual political rights and civil liberties. This is because individuals must deal reciprocally with one another to maintain and perpetuate at the group level, these individual benefits. As democracies are predicated upon individual political and civil rights, they are highly reciprocal in structure. When they encounter similarly constituted states these encounters are characterized by the easy development of multiple, reciprocal, crosscutting ties between a profusion of state (and non-state) actors. This easy growth of dense, reciprocal, ties makes conflict difficult (Axelrod, 1984).

Conversely, interactions between non-democracies, or between a democracy and a non-democracy, would not logically be anticipated to develop such dense, crosscutting, reciprocal, ties. Relatively few inter-state actors would characterize non-democracies, with the state possessing final, summary, decisional authority in all cases. Reciprocity would continue only so long as the ultimate state authority determined it to be useful. Thus all such ties are few, and fragile. War is much more probable in this case, than it is in the latter one.

Culture affects state actor behaviors (Eckstein, 1997). While much cultural variance is nominal, those aspects of culture which determine propensity towards democratic/reciprocal behaviors must, logically, determine, ultimately, a given states propensity towards war-proneness. War-proneness is certainly a non-nominal variable. Thus, culture determines reciprocity, and reciprocity determines war-proneness. War-proneness, in the context of a crisis-driven evolutionary learning system acts as a driver of systemic level learning via the mechanism of adaptation/maladaptation. Adaptive strategies result from the diminishing payoffs attained by war initiators, over time, in the context of historically increasing levels of systemic power.

Increasing systemic power generates the crisis, which the contemporary world system is experiencing. Power derives ultimately, from the amount of information (i.e. "knowledge") available to system actors at any given time. Given that this amount of information has been growing with ever increasing rapidity throughout the approximately 500 year existence of the modern world system, the amount of power available to system actors for inter-actor interactions, such as warfare, has shown a corresponding increase. This increasingly asymptotic rise in the potential for harm to the world system's constituent nation state elements, and consequently, to the viability of the entire system, constitutes the systemic crisis upon which this study is predicated.

Complex Adaptive Systems

I hypothesize, as noted above, that the global system comprises a unitary learning system. Holistic, multi-level, learning systems, as I've articulated above, are, complex adaptive systems. There is no universally accepted definition for a complex adaptive system. A particularly succinct definition is: "A complex adaptive system can be defined as an adaptive network exhibiting aggregate properties that emerge from the local interaction among many agents mutually constituting their own environment." (Tetlock & Belkin, 1996, 260). Still another definition of complex adaptive systems involves, "...a medium-sized number of intelligent, adaptive agents acting on the basis of local information." (Casti, 1996, x). The study's computer model will constitute just such an adaptive network. It will manifest emergent, "bottom-up" properties arising from local interaction among agents (nation-states) which constitute their own environment. There will be a medium sized number of these adaptive agents (10), which will act on the basis of locally available information. Accordingly, the model system will comprise a complex adaptive system by definition. This system will learn to adapt to an endogenously generated selection pressure caused by increasing systemic power, engendering crisis-driven evolutionary learning in the context of dyadic warfare between system elements. To address the question of how the model generated data compare with those of the real world system, we need to look at the behavior of the real world system. Levy's real world findings may be succinctly summarized by his observation that:

> The results are not perfectly congruent across all of these indicators, but some overall patterns emerge. In general, interstate war involving the Great-powers has been diminishing over time. There has been a strong decline in the frequency of war, particularly in the frequency of Great-power war. (Levy, 1983, page 135).

Thus, the real world system appears to be learning how to deal with the crisis, caused by warfare in the context of rapidly increasing systemic power, which threatens its continued existence, by increasingly seeking out non-warfare solutions to inter-state conflicts. In effect, it appears to evince a trajectory through its learning space manifold towards a no-war equilibrium attractor, (a point towards which the apparently random motion of the model system's trajectory converges) driven by selection pressure arising from this endogenously generated crisis. This real world, empirical findings for the 1495 to 1975 period will serve as a basis for empirical falsification of the computer model's theory-derived findings.

Before proceeding, I'd like to note that the propositions advanced above: that, 1) the real world system is adaptive, and moreover, that, 2) it actually 'learns' in the sense that it possesses a 'memory' in which successful adaptations are stored for future reference; are reasonable.

Proposition # 1 just above, must be reasonable, if the world system is to exist at all. After all, this system is characterized by unceasing flux. Its only constant is change. For it to maintain its intrinsic cohesion, given this reality, it must be adaptive. It must possess the ability to adapt to an ever-changing phenomenological reality. No other conclusion is logically possible—other than denying that the system exists at all.

With respect to proposition # 2, Levy's (1983) findings strongly support this thesis. They are consistent with a world system paradigm in which some form of systemic 'memory' plays a major role. For example, his great power observations are highly consistent with the assumption that these powers have access to previous learning regarding the declining efficacy of direct warfare between themselves, in the context of increasing systemic power levels. Further, clear theoretical mechanisms have been articulated for this type of systemic learning (Byron, 1996, 1997). The basic concept is that crisis engenders learning (or war) and this learning, when it occurs, becomes incorporated into the structure of the system in the form of new norms, regimes, and so forth. These serve to enhance the probability that subsequent crises will be resolved via cooperative interaction, leading to new and/or strengthened norms and regimes, rather than by competitive interactions, namely, war.

SECTION II: METHODOLOGY

Construction of the Model

My investigative strategy was to utilize MS QuickBasic to create a model international system program. The model system consists of ten mutually inter-acting actor-elements, which represent the nation-state constituents of the world political system. This value was selected for several reasons. Levy (1983) in his study of great power warfare between 1495 and 1975 identifies a cumulative total of fourteen great powers. Given the assumption that the 10 model system nations represent only those nations which possess sufficient power to significantly influence the trajectory of the entire model system through learning space, and that such model nations are functionally equivalent to Levy's great powers, the model is consistent with phenomenological reality. Also, 10 is a large enough number of model states to allow for collective interactive behaviors to emerge, without being so large a number as to be either too unwieldy to keep track of, or impractical for simulation.

My computer model of crisis-driven learning is predicated, as I've articulated above, upon the hypothesis that systemic level learning occurs principally in response to crisis. In this context, I define crisis as any situation in which the application of previously encoded knowledge fails to resolve some situation in a manner in which harm, or the potential for harm, is averted. In the context of the model, culture "K", represents a common meta-schema for a national polity. Thus, it follows that crisis-induced cultural change will likely occur discontinuously, by discrete increments, which represent replacement of less "fit" schematic elements. This change will occur with maximum probability when it is 'forced', that is, in a crisis situation.

The model consists of four global, "top-down", difference-equation-driven variables along with another variable, which sets an upper limit upon one of these variables. These variables are modified in value via local, "bottom-up" interactions among system actors, by operation of a WAR subroutine, and a CRISIS procedure, in conjunction, where applicable, with the model's TRADE subroutine. Figure 1, below, offers a schematic overview of this model.

As noted above, the four algorithmically defined variables are 1) Relation 2) Reciprocity 3) Power and, 4) Coupling. Each is difference equation defined. As such each imparts global, 'top-down' properties upon the model international system. Appendix One lists these several equations.

a) Relation.

Relation, R: This variable has a range of between +1.0 and -1.0, for each of nation i's relations with other actors comprising the model international system. It represents the instant, net, relationship between any two model systemic nations, "i" and "j". Relation is a zero-sum variable. In effect, each nation possesses a fixed quantity of relational capital. This occurs as the relational difference equation 'assumes' that nations (or more specifically individual/group level nation-state decision-makers) have limited amounts of attentional capacity. Consequently, as more attention (i.e. a stronger signed relationship) is focused upon certain inter-state interactions, the attention available to be focused upon other state interactions diminishes (cf. Jervis, 1976; Simon & Kaplan, 1991; Allport, 1991). For purposes of the model, it is employed primarily to allow for logical consistency. Thus, nations, which engage in warfare with one another, possess a mutually negative interrelationship, which exceeds some threshold value. Conversely, nations, which engage in trade, possess a mutually positive interrelationship, which exceeds another threshold value.

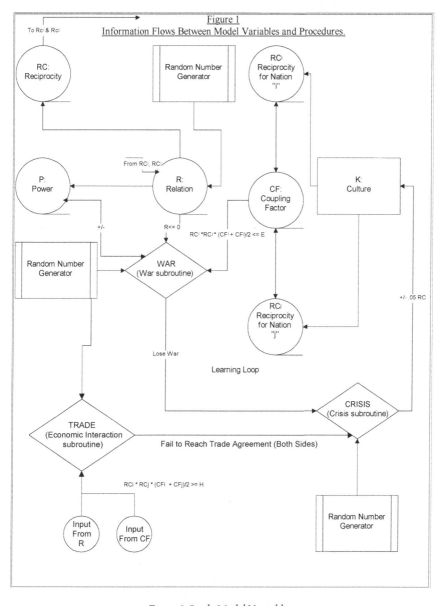

Figure 1: Study Model Variables.

b) Reciprocity.

Reciprocity, RC: This variable represents propensity to engage in reciprocal interactions between any two model systemic nations, "i" and "j". It's the model's indicator for level of inter-democracy lack of war-proneness, in a limited, Axelrodian, sense.

According to my theory of crisis-driven learning, increasing levels of systemic learning are characterized by an increasing profusion of international regimes, in the context of an ever broader and deeper diffusion of international norms. This has the effect of increasing mean values of systemic reciprocity. Therefore, mean systemic reciprocity will serve as a measure of net systemic learning for any arbitrary time "t."

c) Power.

Power, P: This variable represents total power which a given state, "i", possesses at any time t. Power, in the context of the model, may be defined as the ability of a given state "i" to potentially influence, or affect, the behavior of some other state "j", while possessing a commensurate ability to resist being involuntarily influenced by state "j".

The power equation 'assumes' that propensity towards reciprocal behavior is conducive to increasing a given state "i's" relative share of total systemic power. Conversely, non-reciprocal behavior decreases "i's" relative power. My reasoning is that cooperation allows limited resources for state "i" to be 'productively' utilized. Low reciprocity states, on the other hand, choose to 'squander' some portion of their power by engaging in conflict. This (culturally determined) behavioral strategy becomes increasingly less likely to yield net benefits (i.e. an increase in the power of war-initiator states) as systemic power increases ever more rapidly. Thus, their relative share of total systemic power tends to diminish, on average, over time. This ultimately leads to increased 'selection pressure' upon them.

The model is predicated upon an assumption that net systemic power is increasing over time, (as a function of, generally available, increasing knowledge) concomitantly generating a crisis of system survival, in the context of a war dominated international system. Consistent with this assumption, the power difference equation incorporates a term for incrementing power from program iteration to program iteration, somewhat like compound interest.

d) Culture.

A variable, "K" represents culture. K has a potential range of 0.00 to 1.00. initial values of K are arbitrarily assigned with 5 nations receiving "high bound" culture K values of .75, and the other 5 receiving "low bound" culture values of .25. This yields a mean initial K value of exactly .5. I selected this value in accord with the explicit assumption that it represents the lowest mean K value at which system coalescence, in a manner analogous to that of the modern world system circa 1500 AD can occur. K simply sets an upper (though not a lower) limit for the potential range of state "i's" RC values. RC values are free to vary within the upper bound set by this limit. K varies by discreet increments of .05 because cognitive, or behavioral, change is assumed to occur discretely (i.e. by schematic and memetic replacement) as I've discussed above.

e) Coupling.

Finally, the model incorporates a coupling parameter, "CF". CF provides a measure of how closely coupled any two nations "i" and "j" are at any time "t". In effect, it determines the extent to which any nation "i" may interact with, and thus algorithmically influence, the behavior of a given other nation, "j". Evaluated for the system as a whole (by computing its mean value) it provides a measure of the degree of "openness" of the system. As such it is closely related to reciprocity, as a perfectly reciprocal system would be anticipated to be a highly coupled system. Conversely, a perfectly non-coupled system would be anticipated to exhibit little, if any, reciprocity. Thus, reciprocal systems would be expected to exhibit numerous national and sub-national, bi-directional interaction or communications channels. Correspondingly, these relational pathways would be expected to decline in tandem with diminishing coupling. Its difference equation "assumes" that perfect coupling can be closely approached, but never actually reached.

Taken together, these four difference equation determined variables provide the model system with a coherent "top-down" structure. That is, they constitute global, deterministic, transformation procedures that determine how the model system evolves with respect to system time.

Study Model Subroutines.

a) WAR Subroutine.

The model's WAR subroutine is triggered when the following criteria are met: IF $RC_{ij}(t) * ((CF_i + CF_j) / 2) \leq E$ AND $R_{ij}(t) \leq 0$ THEN GOSUB (i.e. go to subroutine) WAR. "E" is an algorithmically determined positive, triggering threshold value that will be described in a subsequent section of this paper. The WAR subroutine produces, in most cases, a "winner", as well as a "loser." Winners have their power increased, while losers suffer a corresponding decrease. About one out of ten times no winner/loser is selected. This reflects the ambiguities inherent to warfare. Losers also undergo the CRISIS procedure. This subroutine provides a "bottom up" stochastic, process which allows for non-deterministic evolution and learning to occur and influence the trajectory of the model system. My thinking here is that warfare "winners" having successfully increased their power via application of existent behavioral algorithms (represented by their culture or "K" value) will experience little, or no, pressure to modify their behavior. Conversely, the application of these existent behavioral algorithms has resulted in grievous harm, in the form of diminution of power, to the war loser. Per my theory of crisis-driven learning, this situation ought to constitute a crisis, as I've defined it above. This actor perceived crisis, per theoretic assumptions, leads to an adjustment of the actor's behavioral algorithms.

b) CRISIS subroutine.

Nations which lose at WAR experience two outcomes, the first, an algorithmic reduction in their power, was mentioned above. The second is that they undergo the CRISIS procedure. Here, their K values are randomly raised, or lowered, by an increment of .05. The probabilities for each outcome are equal. This discrete value is selected in conformity with the model's assumption that cultural change occurs adaptively/maladaptively, by discrete intervals, corresponding to schematic/memetic replacement, of less "fit" behavioral schemas in response to crisis. Continuing inter-state warfare in the context of increasing systemic power, as discussed above, engenders crisis.

In essence, via operation of the model's WAR subroutine, and its CRISIS procedure, each nation, sequentially, compares itself dyadically, with all other nations. This comparison determines whether the two nations will engage in war, and if so, which one will initiate it. The effects of this process include substantial adjustments in power and, for losers at war, culture, and hence, reciprocity. It is this local, interactive process, which allow for the system's adaptive learning behavior to emerge locally within the context of its difference equation specified global "top down" structure. This subroutine allows for additional, non-algorithmic, "bottom up" learning to take place in the context of non-warfare interaction among systemic actors. It is here, in the context of the model, that adaptation occurs. The result is an additional, stochastic, input, affecting the model system's trajectory through learning space.

c) TRADE Subroutine.

This subroutine operates similarly to that of the WAR subroutine. Its triggering threshold is represented by the variable "H." The principle differences are that relations between nations "i" and "j" must be positive, (i.e. RCij) as opposed to negative, for the WAR subroutine. The triggering values ("H"), as we shall see below, are, as is the case for the WAR subroutine, an algorithmically determined variable. Also, for purposes of this inquiry, if failure to reach a trade agreement occurs, BOTH actors undergo the CRISIS procedure, and hence learning, as opposed to only the loser doing so in the in WAR subroutine.

My reasoning is that if crisis is modeled to be the primary driver of systemic learning, then any class of conditions, the failure of which would likely be perceived by system actors as creating harm to their well being, ought to be a generator of crisis. This is because of the fact that as both actors are similarly deprived of their potential gains both similarly experience crisis. Simply put, both, by being deprived of their potential winnings, become "losers". This triggers the CRISIS module.

I should note that the CRISIS procedure is algorithmically triggered in this manner with between about one-fifth, to one-tenth, the frequency which it is triggered by the corresponding algorithm located in the WAR subroutine. Both algorithms are in part deterministic, relying upon instant values of system variables, and in part stochastic, relying upon input from a random number generator. For high levels of reciprocity trade-instigated CRISIS decreases rapidly, diminishing to zero at very high RC levels. This reflects my assumption that crises are generally perceived by state actors as offering less potential for harm than is the case for defeat in war.

The TRADE subroutine is predicated upon the explicit assumption that all trade interactions are mutually beneficial, (recall the discussion above concerning Ricardo's Law of Comparative Advantage) in terms of increased power accruing to each party. At relatively low levels of mutual reciprocity (RC^{ij}) these benefits accrue disproportionately to the more powerful of the two trading entities as a function of the ratio of their respective power. However, as RC^{ij} increases, the distribution of benefits becomes progressively more equitable. For high values of RC^{ij} (as I shall explicate just below) it is actually *possible* for the less powerful nation to benefit disproportionately. In so doing, the TRADE algorithm reflects my interpretation of the application of Ricardo's Law of

COMPARATIVE ADVANTAGE TO THIS CLASS OF INTERACTIONS.

Study Model Parameters

Operation of this theory-predicated model creates a learning space manifold in which systemic learning occurs. Evaluation of system behavior in the context of this manifold requires the introduction of several new systemic measures.

The first of these is termed "PI." PI represents the value of the ratio of instant systemic power to initial systemic power. It provides an indicator of the model system's absolute magnitude of systemic power increase. As there is a very great range within which power may vary between the inception and termination of each model system, the resultant PI values for two or more model system's can vary by many orders of magnitude.

A second is "C". C varies uniquely for each system. It has a maximum value of *approximately* 1.0, (each model system is unique) and a minimum value of 0.0. It provides an indicator of the model system's trajectory within the learning space matrix. It does this because instant systemic power and systemic position within the manifold are complementary concepts. It is calculated according to the formula: $C = |2 * (1/ P_{total}{}^{-2})|$. Because it is logarithmically derived from the instant value of systemic power, it provides an (logarithmically scaled) indication of systemic 'depth.' This scaling methodology allows for easy visualization of changes in systemic power. As these changes can occur over many orders of magnitude, direct viewing is wholly infeasible. When C is evaluated with respect to system time, it provides an indicator for systemic trajectory. Were it to be displayed three-dimensionally, it would offer a glimpse of the topology of the learning space manifold itself.

Yet another parameter is labeled "CYCLE". It designates the number of program iterations, which have transpired. As such it provides the measure for system time. By definition CYCLE represents the time in which one iteration of the model system program occurs.

There is also a nominal distance parameter, "D." It possesses a potential range between approaching zero, to approaching one. It is lowest for "far apart", low net reciprocity, nations. Conversely, it is greatest for "nearby" high reciprocity nations. "Near" and "far" in this context are nominally determined by the absolute value of the difference between each model nation's pre-assigned identity number. The ten model nations are arbitrarily numbered as nations 1 through nation 10. Thus, nation 1 and nation 5 are separated by an absolute value of 4, for example. This is their "distance." As this distance may be 'bridged' with varying degrees of efficiency depending upon internal conditions within nations i and j, its *effective* value is modified somewhat by their unique, instant, values of reciprocity. I assume that more highly reciprocal nations are better able to 'bridge' distances.

D is utilized, in conjunction with reciprocity, in determining which nation in a given ij dyad will "initiate" trade or war. This is significant as initiators of both trade and war are treated differently, in terms of calculating power for the given dyadic interaction, than are non-initiators. This asymmetry tends to "favor" the initiators of trade, while correspondingly "penalizing" (in terms of net power change arising from the transaction) the initiators of war. The magnitude of this biasing ranges from negligible/nonexistent at very low levels of systemic power, to being very pronounced for high levels of systemic power. It is also utilized in calculating the relative power which each member of the ij dyad brings into the trade or war interaction. Overall, D serves to allow the model's assumption that continuing warfare, in the context of rapidly increasing systemic power represents a threat to the existence of the system, along with the well established corollary to this that trade and warfare are inversely related (e.g. Mansfield 1994), to incorporate nominal distances between model nations into the model's ij dyadic relationship calculations. Simply put, the farther 'apart' any two nations are, *ceteris paribus,* the less power that they can bring to bear for any mutual interaction.

My model system program operates by sequentially and dyadically evaluating the relationships which exist between a given model nation "i", and the other nine model nations. This includes determining whether or not "i" should engage in war or trade interactions with another nation "j", and if so, whether stochastic interaction factors determine that it undergo crisis-driven learning. Once all 10 nations have undergone this process, the program repeats these procedures, thus beginning a new program iteration, or cycle.

The *TRADE subroutine also incorporates a systemic metric parameter "G."* This is defined as being 7.75 – total systemic reciprocity. As total Rc can possess any value within the range 0.0 to 10.0, this systemic parameter decreases in value as a given model system becomes more interactively reciprocal. For very high Rc systems, it actually becomes negative. The effect of G is to lessen the structural inequality which otherwise exists between relatively more and less powerful trading nations. For very high values of Rc, it allows for substantially less powerful nations to, potentially, benefit in an absolute sense from trading interactions. My reasoning here is that high reciprocity systems are qualitatively different from low reciprocity ones. This is because between group cooperative interactions (trade) come increasingly to be favored over between group competitive interactions (war). This has the effect of increasingly reorganizing the structure of the international system in such a way as to 'learn' to incorporate Ricardo's Law of Comparative Advantage into its structure. As these effects do not exist for a power-based system, G is utilized only in the TRADE subroutine. When G is negative, trading interactions may constitute an actual transfer of power (wealth), in an absolute sense, between a more, and a less powerful, ij dyad. This occurs not due to an unrealistic assumption of altruism, but rather because once the "rules" of the system come to fully incorporate the law of comparative advantage, asymmetric power based relationships become relatively less beneficial to the longer term interests of both parties. Both will benefit more if their interactions occur on the basis of relative, functional equality. Thus my reasoning here simply operationalizes the logic of increasingly integrating the law of comparative advantage into the structure of the international system via the mechanism of crisis-driven evolutionary learning.

The learning space manifold within which these various interactions occur is configured as a 10-dimensional array. Each "dimension" corresponds uniquely with one of the model's 10 constituent nation states. Each time the program is initiated a unique model system is created de novo and positioned in an arbitrary location in this space, consistent with a moderately low level of learning.

Recalling that each of the 10 dimensions is, in actuality, a range of discrete values provides direction in this space. Motion along this range towards a no-war equilibrium solution (that is towards an attractor which represents a discrete solution to the problem of attaining this condition) corresponds to motion along the C-axis towards its end-point, 0. Thus, lower C values, and correspondingly, higher amounts of learning, (a closer approach to a no-war equilibrium problem attractor- solution) correspond to motion "deeper" into the learning array.

This visualization also allows for the geometry of the learning array to be comprehended: As the end-point, or vertex, for C is common to all 10 nations, then this vertex is the common vertex of learning space. All 10 dimensions therefore meet at this one point. At this unique point, and only at this point, (which corresponds to perfect systemic reciprocity) they coincide.

As I visualize it, the learning space manifold looks somewhat like an inverted 10-sided step-polygon, which steps 'downwards' towards this singularity in increments of .05 K. Each of these points is located exactly .05 K units from all adjacent points. Its 10 sides, or dimensions, as noted above, correspond to the 10 nations that comprise the model system. Topographically, the model's learning space, which lies within the interior volume of this 10-sided inverted step-polygon, is contorted by the presence of attractors, which represent unique solutions to a system's no-war, crisis-driven, equilibrium search.

It is important to note that this learning space manifold is *endogenously generated*. This means that the learning space manifold is not some pre-existent metric, into which a model system is placed; rather it exists because a given system exists. Within the top-down parameters allowed by the modeling software, it is free to vary as a function of instant, unique, systemic characteristics. Thus, even the 'location' of attractors *can* change as a given system evolves.

A final systemic parameter is generated by operation of the collapse subroutine. This module algorithmically determines the probability that a warfare-prone system will collapse as a result of continuing warfare in the context of rapidly rising power levels. As there is insufficient empirical data to determine actual, real world historical probabilities, its values are, necessarily, somewhat arbitrary. However, they do provide a standard metric against which the effects of varying one or more systemic parameters can be evaluated. Succinctly, the probability of systemic collapse is very low at low levels of systemic power. As power increases, so does the probability of collapse. For very high levels of systemic power it becomes unity. The collapse probability increases occur discontinuously; as a given power threshold is attained, the collapse probability rises correspondingly, in a stepwise manner. At this point we are now able to proceed with the analytical portion of the study.

SECTION III: ANALYSIS AND EVALUATION OF THE MODEL

The study's hypothesis is that the international system constitutes an adaptive learning system. It intends to determine whether this learning system is best characterized as being dominated by between-group conflict (realism-type paradigm, or whether between-group cooperation exerts a significant influence over systemic processes (liberalism-type paradigm). To begin to evaluate these propositions I created five variants of the model. The first of these referred to as "PWar" is identical to the next three except that it does not utilize the model's

TRADE module. It corresponds to a pure, realist (e.g. Gilpin, 1981) international system paradigm. As such it is the control, or baseline, systemic model against which the other variants are to be compared.

Subsequently, I generated three variants of the model, which were identical to the above-described "PWar" system, except that they employed the TRADE module. These model systems utilized a fixed threshold value to determine algorithmically whether, or not, conditions were appropriate for a given nation "i" to engage in trade with another nation "j". The purpose of this is to investigate the sensitivity of the outcomes for trading interactions to arbitrarily preset parameters.

These three model systems are labeled: 1) "hEWar" (i.e. "high TRADE module enabled War), 2) "mEWar" (where "m" represents "medium"), 3) "lEWar" ("l" for "low"). They meet the conditions for triggering the TRADE subroutine (and hence international trade) specified by: IF $RC^{ij}_{(t)} * ((CF^i + CF^j) / 2) \geq H$ AND $R^{ij}_{(t)} > 0$ THEN GOSUB TRADE, where H is equal to .35, .30 and .25 respectively.

In the real world system there is no analog for these fixed threshold values. It is a self-regulating, complex adaptive system, as discussed above. Accordingly, I created a fifth, and final, version of the model, termed "ArWar" (auto-Regulating War). This class of model system algorithmically calculates its own triggering threshold values in accordance with a simple formula based upon the instant systemic value of reciprocity (RC): IF $RC^{ij}_{(t)} * ((CF^i + CF^j) / 2) \geq (1 - (.85 * \text{MEANRC}))$ AND $R^{ij}_{(t)} > 0$ THEN GOSUB TRADE. Here the term $(1 - (.85 * \text{MEANRC}))$ replaces the fixed constant "H" used for the preceding three system variants. "MEANRC" designates the instant value of mean systemic reciprocity.

Thus, the instant values of the TRADE module's triggering threshold depends critically upon the temporally corresponding values of mean systemic reciprocity. This is *non-arbitrarily* consistent with the model's underlying assumption that levels of international trade are dependent upon the relative absence of warfare from the international system: more warfare engenders less trade, until a threshold is reached at which trade ceases. As, in the context of the model, reciprocity and war are inversely correlated, it follows logically that a system characterized by higher levels of reciprocity ought to be characterized by higher levels of international trade.

The simple $1 - (.85 * \text{MEANRC})$ threshold-setting algorithm serves to maintain logical consistency with core theoretic assumptions. All model systems begin with an initial mean systemic RC of .5, thus, the initial TRADE threshold is .575. This means that the early model international system is highly anarchic, and war-prone, making trade difficult. As crisis-driven learning occurs the system gradually becomes less anarchic, and hence more conducive, for trade. Thus, the triggering threshold for trade must evince a corresponding decrease. At a mean systemic RC level of approximately .882 the triggering threshold for international trade becomes equal (i.e. .25) to that for interstate war. At higher levels of mean systemic reciprocity it actually becomes very slightly less. Relatively few model system runs attain such high levels of mean systemic reciprocity without either reaching a no-war equilibrium, (undergoing a cessation of inter-state warfare) or collapsing due to continuing warfare. However, those that do are generally characterized by the presence of one, or two, states that still engage in inter-state war, while being surrounded by an otherwise peaceful system of trading states. In such a systemic configuration, and only in such a configuration, the model allows two states, which go to war, to also engage in trade, during the same program cycle. The reasoning here is that this type of system has become predominately "trade-like", as opposed to "war-like". Accordingly, it evinces different behaviors when stressed. Looked at from another perspective, trade could be considered to be a "weapon" to integrate an aberrant member of the otherwise pacific international system.

I ran each of these five systemic variants 500 times in order to produce a statistically meaningful $\underline{N}$ of cases. The resulting data files were read into EXCEL, and sorted by whether, or not, they had collapsed before inter-state warfare had ceased. Subsequently, I used the system variant with the fewest valid runs, (PWar) with 347 valid, or non-collapsed, cases, to establish the $\underline{N}$ for the study. All valid cases numbered above 347 for the other four system variants were discarded. Elimination of these excess cases was purely random, involving no bias on the part of the author, other than accepting Excel's numbering of them arising from the above described sorting procedure. These data are displayed in table 1 below.

System$\underline{N}$	Cycle	PI	C	RC	CF	Collapse%	
PWar	347	92.91	273,255.91	.245	.816	.773	.306
hEWar	347	56.78	556,130.05	.247	.878	.832	.272
mEWar	347	33.52	196,902.36	.278	.885	.838	.276
lEWar	347	24.94	183,860.78	.316	.883	.836	.140
ArWar	347	86.64	123,687.03	.246	.859	.814	.194

Table 1. Systemic Data.

It is clear that selection of any of the several fixed values for triggering the TRADE module produce systemic outcomes that are 'preferable' to those of the baseline PWar type system in that the probability of collapse is lower, while the corresponding values of systemic reciprocity and coupling are greater. Values for C do tend to be greater for the fixed TRADE subroutine triggered cases. However, as learning continues via trade interactions for this class of system, this distinction is not critical [All comparisons are significant at the .05 level, or better.]

Of course, it is not intuitively reasonable to assume that trade interactions in the real world system possess a fixed triggering threshold. The ArWar model systems avoid this difficulty, as described above. A quick glance at figure one should serve to make clear that ArWar type systems are substantially more adaptable, overall than are any of the other system variants. Terminal C values for ArWar and PWar systems did not evince a statistically significant difference.

Overall, it is clear that systems, which are influenced by economic interactions, learn with greater rapidity. This conclusion is buttressed by noting that all trading systems studied were able to seek out no-war equilibrium attractor-solutions more rapidly than was the case for the non-trade dominated systems. In all cases, their probability of systemic collapse was substantially reduced. Reciprocity was greater for this class of system. Additionally, they were substantially more highly coupled. Learning occurred at much lower levels of net mean systemic power (PI) than was the case for the realist paradigm PWar systems. No statistically significant difference was found between the mean terminal C values of the two classes of systems. Each is thus equally 'stable', ceteris paribus. However, as noted in the discussion above, learning continues after the cessation of inter-state warfare via the mechanism of interstate trade crises. Thus, even in this category, the trading systems exhibit greater viability. As noted above, this provides compelling tribute to the power of Ricardo's Law of Comparative Advantage. Given these findings, I incorporated the ArWar TRADE algorithm in to the baseline model. Doing so, in the context of the above assessment, is a-priori reasonable in that it brings the model into closer correspondence with phenomenological reality, is consistent with crisis-driven evolutionary learning (as it renders the system more 'adaptive'), and does not represent arbitrarily selecting some optimal value in order to make the model 'work.'

Building upon this, I next altered the manner in which the WAR sub-routine's triggering algorithm is calculated. Initially, this value had been a simple constant value of .25. This value has been chosen as it was intuitively reasonable. Recall the WAR subroutine triggering algorithm: IF $RC^{ij}_{(t)}$ * ((CF^i + CF^j) / 2) ≤ .25 AND $R^{ij}_{(t)}$ ≤ 0 THEN GOSUB WAR. Thus, any two nations "i" and "j" which possessed a *mean* reciprocity of .65, or *less*, and which had a signed relational value which was *negative*, could, potentially, go to war with one another. This corresponds with the real world observations that democracies do not go to war with one another as two high reciprocity (RC values of above .65 per algorithmic definition) model nations would not meet the dyadic war criteria. Consistent with this, two low RC nations would potentially, meet these criteria. Furthermore, as per empirical observation, war between a "democratic" (defined as a model system nation possessing a RC value in excess of .65) and a "non-democratic" model system nation would be, potentially, possible. The problem with this fixed threshold is that it is predicated upon an unrealistic assumption: that wars occur, in all times, and under all circumstances, under invariant conditions. Indeed, if this assumption were to be true, then learning derived-changes in the structure of the international system would have no effect upon the conditions under which wars occur, only upon their frequency. It appears much more logical to me to posit that the threshold for war occurrence is lower when the international system is more anarchic, and that, conversely, it is significantly greater when the system is characterized by trading interactions among high reciprocity nations.

To operationalize this assumption, while preserving the intuitive and observationally derived warfare assumptions embodied by the .25 war-triggering threshold value, I developed a simple algorithm to produce this dynamic triggering value: $RC - (/1.65)$. This dynamic triggering algorithm produced a *mean* triggering value of very nearly .25, thus preserving the benefits of the fixed value threshold, yet it varied as a function of mean RC, thus allowing warfare triggering conditions to vary in an intuitively reasonable manner. For example, in a 500 $\underline{N}$ run test, its value, averaged over the 500 runs, was .2544 for the PWar class of model systems, and .2489 for the ArWar class of model system.

Figure 2
Trade and War Triggering Thresholds.

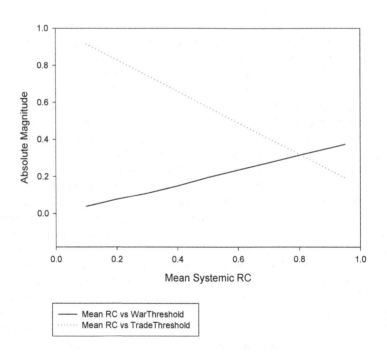

Mean RC vs WarThreshold
Mean RC vs TradeThreshold

Table 2. Selected Systemic Measures for Varying WAR-triggering Threshold Values.

Figure 2 below displays the resulting slope of the WAR-triggering function, in conjunction with that for the above-discussed TRADE-triggering function.

Note that: A) The slope for the TRADE-triggering algorithm is *steeper* than is that for the WAR-triggering algorithm. This depicts the operationalization of my assumption that trade can, generally, occur more easily, and under a broader range of systemic conditions, than can war. B) At a value of mean systemic RC of about .81 the trade and war algorithm slopes intersect. For higher mean systemic values than this, a given "ij" dyad can engage in *both* war and trade during a *single* program cycle. I explain this by noting that for high RC systems, war, instigated by one, or two, low RC nations, is an anomaly. Under these conditions, for such a "tradelike" system, trading with war-prone nations becomes a type of defensive "weapon" designed to promote the integration of the rouge nation into the otherwise pacific, tradelike, international system. Table 2 below presents the data for the relationship between various fixed values of the war-triggering threshold and several systemic variables and parameters. For comparison, corresponding values for ArWar and PWar systems are also provided. These systems are fully self-regulating with respect to determining their war and trade thresholds. The methodology employed is identical to that utilized above in Table 1, except that an N of 604 cases were required (rather than 500) to obtain the 226 non-collapsed cases for the .50 triggering threshold data. This is because I made a decision that a study N of at least 200 cases would be required to allow for significance testing. An unadjusted N of 100 cases was generated for triggering values of .60 and .70. This was done as their collapse rates exceeded 90 percent. Consequently, these data must be considered as being less reliable than the rest of the data in Table 2.

Triggering Threshold	N	Cycle	Collapse%	PI	C	RC	CF
.10	226	47.33	0.4	156.2	.520	.686	.650
.20	226	75.22	8.4	179,910.9	.314	.810	.768
.25	226	83.01	14.8	105,089.8	.261	.867	.822
.30	226	88.80	22.0	88,493.6	.216	.894	.847
.40	226	118.79	39.9	404,726.7	.176	.915	.867
.50	226	154.62	61.9	75,038,198.6	.130	.939	.793
.60	100	211.85	93.0	161,426,525.0	.105	.837	.793
.70	100	270.63	98.0	4,449,232.5	.100	.812	.792
PWar	244	109.57	51.2	338,257.9	.188	.836	.792
ArWar	244	92.29	14.8	67,952.1	.227	.891	.845

Per table 2 the systemic parameters of Cycle, Collapse, and C, along with (probably) PI, vary linearly with respect to value of WAR subroutine triggering threshold. Conversely, RC and CF evince a clear, curvilinear, variance with respect to this threshold. Their values are maximal between triggering values of .40 and .50. As explicated above the WAR triggering mechanism is *intended* to produce a mean threshold value of around .25.This is predicated logical grounds, in the context of my crisis-driven evolutionary learning theory. As this value is less than .40-.50, this selection may appear substandard. However, I would note that we can now see from table 2 that this lower value also corresponds to lower values for Cycle, and Collapse. Hence when the totalities of all systemic parameters are taken into account, this lower value appears to more nearly represent the optimal triggering value. Which of these systemic classes best correspond to real world data? Recall Levy's (1983) comments:

> .... some overall patterns emerge. In general, interstate war involving the Great-powers has been diminishing over time. There has been a strong decline in the frequency of war, particularly in the frequency of Great-power war. (Levy, 1983, 135).

Clearly, in both classes of model system inter-state warfare is observed to decline with respect to system time. What about declines in the relative frequencies of war for 'great powers' vis-à-vis 'minor powers'?

Figures 3a through d display the relative warfare frequencies for representative power-dominated ("PWar") and trade enabled ("ArWar") model systems. They are drawn from the N of cases utilized to produce the PWar and ArWar data displayed in table 1 above. They were selected on the basis of their being as nearly representative of the entire set of 244 cases utilized, as possible. What each figure vividly depicts is a decrease in warfare frequencies, both absolutely, and as a function of increasing power.

It must be noted that the mean terminal systemic values presented in table 1 above are statistical composites. No given system actually possesses these values for more than one, or possibly two, of its several terminal systemic parameter values. The below data were selected with this in mind. These data are presented below in table 3.

Table 3. Terminal Systemic Parameters.

System #	CYCLE	PI	C	RC	CF
PWar 1	90	136,433	.11	.80	.75
PWar 2	91	6,015	.16	.82	.78
ArWar 1	82	1,245	.22	.88	.83
ArWar 2	88	490,064	.07	.90	.85

Following Levy's (1985) operationalization procedure, model system nations were grouped into two categories: great powers and minor powers. Any nation which possessed at least 15% of total systemic power during a given

Figure 3a
Relative Warfare Frequencies for # 1 Power-Dominated Model System.

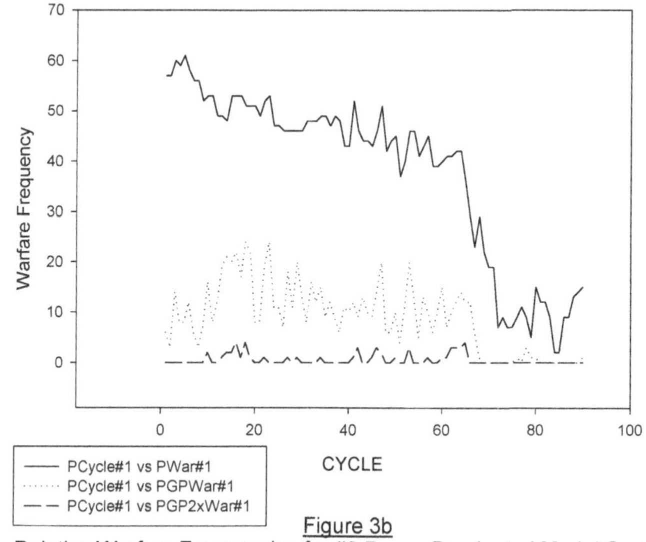

Figure 3b
Relative Warfare Frequencies for #2 Power-Dominated Model System.

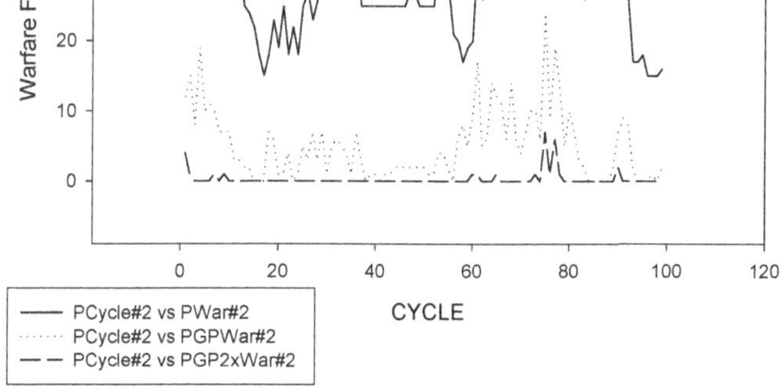

program cycle was assigned to the great power category for that cycle. As relative power between model nations changed, membership in these groupings

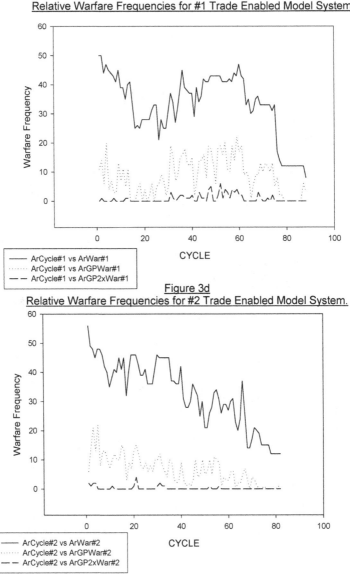

Figure 3c
Relative Warfare Frequencies for #1 Trade Enabled Model System.

Figure 3d
Relative Warfare Frequencies for #2 Trade Enabled Model System.

also varied. The three categories employed in figure 2 above represent, respectively, all wars (Wars), wars involving at least one great power (GPWar), and wars between great powers (GP2x). Per Levy's real world empirical findings, we would expect wars between great powers to decline most precipitously, with respect to system time, wars involving at least one great power to decline somewhat less rapidly, with all wars, including minor power vs. minor power wars to decline with greater relative slowness. A glance at figures 3a through 3d indicates that this pattern is clearly manifested for both classes of model systems. In both cases, those of the power-driven model systems depicted in figures 3a and 3b, as well as for the representative trade-enabled systems, figures 3c and 3d, warfare frequencies are found to decrease, on average, over time, both absolutely, and as a function of increasing nation-state power.

Thus Levy's findings fail to falsify either hypothetical systemic configuration. Nonetheless, as both model systems are variants of a complex adaptive learning systemic model, their close correspondence with Levy's real world empirical findings does evince strong support for the core hypothesis that the real world is indeed such a learning system. The remaining issue concerns the role of economic interactions in determining the temporal evolution of the actual world system.

At this point it will be informative to evaluate the trajectories and topologies of representative examples of each class of sample model system. Figures 4a through 4h accomplish this. The set of model systems referenced in table 3, above, are utilized for this purpose. Sigma Plot's 3-d modeling software was employed to create these graphs.

This type of data display portrayed by figures 4b, 4d, 4f, and 4h, below, is knows as a "mesh plot." System time and warfare frequencies form its x and y-axes. These two dimensions, in effect display the area of the manifold in which inter-state warfare—as well as systemic learning as to how to avoid war—occurs.

Depth, in the form of a z-axis, is provided by the systemic measure "C" as described above. By recalling that C provides an (logarithmically adjusted with a range between *approximately* 1.0 and 0.0) indicator for the model system's instant position, or "depth" in the learning space manifold, the rationale for its selection as the z-axis, or depth information variable, should be clear.

Thus, each mesh plot graph depicts third-dimensional, topological contortions, of a 2-dimensional surface, which represents warfare frequencies with respect to system time. Recall that each model system is experiencing endogenously generated selection pressures, in the context of increasing systemic power, to find a solution to its crisis of continuing warfare. Also recall that motion 'deeper' into the learning space manifold corresponds to greater systemic 'learning' (in the form of higher mean K and RC values, for example). Now, because this learning space is endogenously generated, with each individual model system being unique, it must follow that there are unique locations, or points within this manifold which correspond to precisely those systemic values which a given system requires in order to 'learn' to eliminate inter-state warfare. The deeper such a point lies in the manifold, the more 'stable' it is. [Think of a

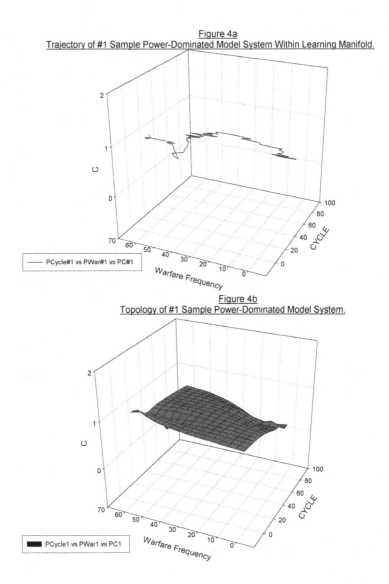

Figure 4b
Topology of #1 Sample Power-Dominated Model System.

marble rolling into a shallow depression—it can easily roll out again. As the depression becomes deeper, this outcome becomes progressively less probable.]

Putting all of this together results in the mesh plot graphs being able to be interpreted as representing *fitness landscapes* for each model system. In the context of these landscapes, increasing *depth* corresponds to increasing *fitness* (where fitness represents decreasing war-proneness). The depressions can now be seen

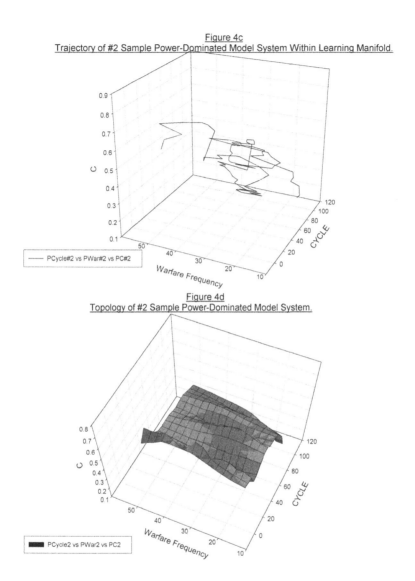

Figure 4c
Trajectory of #2 Sample Power-Dominated Model System Within Learning Manifold.

Figure 4d
Topology of #2 Sample Power-Dominated Model System.

as representing the location of no-war equilibrium *attractors*. Deeper attractors are more stable than those located higher up on this fitness landscape. Thus the system can be considered as 'wandering' randomly across this fitness landscape, driven by selection pressure to seek out a *stable* attractor, which represents a valid solution to its no-war equilibrium problem. Figures 4a, 4c, 4e, and 4g, represent exactly this. As such, they are 3-d graphs of their respective system's trajectory across the corresponding fitness landscape. Taken together, each set of

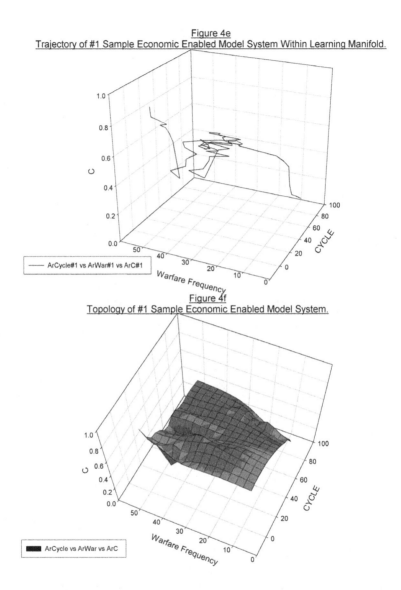

figures graphically represents both the trajectory of each model system within its learning space manifold, as well as the contortions of that learning space manifold which are indicative of the presence of these point attractor-solutions.

Note that power-dominated systems *on average* evince lower dimensionality, and hence appear as being more nearly 2-dimensional in figures 4a through 4d. Conversely, the economic enabled model systems *generally* evince substan-

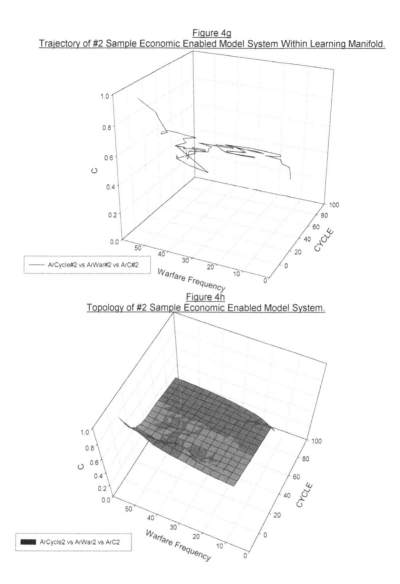

Figure 4g
Trajectory of #2 Sample Economic Enabled Model System Within Learning Manifold.

Figure 4h
Topology of #2 Sample Economic Enabled Model System.

tially higher dimensionality, and thus appear to be fully 3-dimensional in figures 4e through 4h. This indicates that the trade-enabled systems are significantly more volatile. They adapt more often, and with greater rapidity, on average, than do the power-dominated systems.

Empirical corroboration of this observation is provided by recalling that "depth" information is provided by the system parameter "C." If the hypothesis that ArWar systems, exhibit, *on average*, greater dimensionality (technically, greater *fractal* dimensionality) than do corresponding PWar systems, were valid, then I would logically anticipate that corresponding values of C for each model system variant would also exhibit relative differences in dimensionality. To test this hypothesis, I evaluated an N of 25,969 consecutive cyclic values of C for each systemic variant. Mean kurtosis values for ArWar systems were found to be -.319. The corresponding values for PWar systems were -.394. These differences were statistically significant at better than the .001 level. A second test of the hypothesis utilizing an N of 24,774 other, randomly generated cases, yielded corresponding values of -.248 and -.439, respectively. Kurtosis is a measure for how much a given distribution varies from normal. Distributions, which are negative, are termed *platykurtic*. Such distributions are, on average, "flatter" than is a standard normal distribution. Both the Ar and P distributions fall into this category. However, the Ar distribution is substantially less "flat" than is the P distribution. That is, it exhibits substantially higher (relative) dimensionality than does the P distribution. This finding is exactly what would be logically anticipated to occur if the hypothesis that ArWar systems show greater fractal dimensionality, and hence volatility, and hence adaptiveness, than do PWar systems were, indeed, true.

SECTION IV: DISCUSSION

Clearly, both types of model systems produce similar outcomes with respect to Levy's empirically observed real world phenomena of warfare frequencies, which decline as both a function of time and power. This is almost certainly due to their both seeking out no-war attractor-solutions in their endogenously generated learning space manifolds.

Equally clearly, the trading systemic variants exhibit higher dimensionality. Corollary with this, they exhibit greater (and more rapid) adaptiveness. My conclusion is that this visually represents the operationalization of the law of comparative advantage upon this class of system's trajectory and topology. Recall Ridley's (1996, pp. 210) comments:

> ....There simply is no other animal that exploits the law of comparative advantage between groups. Within groups, as we have seen, the division of labor is beautifully exploited by the ants, the mole rats, the Huia birds. But not between-groups...The Law of Comparative Advantage is one of the ecological aces that our species holds.

This is the source of the volatility, the adaptiveness, the dimensionality, observed above. So which systemic variant best corresponds to the real world system? I think that the question is misleading. Here's why: Both system variants are crisis-driven, complex adaptive learning systems. As such, both operationalize in such a manner as to produce results consistent with corresponding real world empirical data.

Given these conclusions, it follows that such a world system must become more complex, in the sense that it incorporates ever more information into the system's structure (Byron, 1997; Modelski, 1987). This incorporation takes the form of a proliferation of international regimes, global and regional free-trade agreements, software piracy "understandings", new norms such as nuclear non-proliferation, communications agreements, military alliances refashioned to protect against "chaos" (i.e. NATO), and so forth. Reciprocal trading, informational, decisional, and communications pathways multiply. Epistemic communities begin to flourish (e.g. Haas, 1990).

What is happening is that the global system is becoming ever more sensitive to violent disruption, ever more vulnerable, to serious harm, or to actual collapse. In tandem with this trend, the amount of military power available to inflict just such harm is growing exponentially. Simultaneous with these trends, the global system strives to maintain its coherent existence. This creates circumstances in which rapid adaptation, which incorporates a 'memory' for previous learning, is essential to systemic survival. If there were ever a time, place and circumstances in which crisis-driven learning would be anticipated to occur, this would surely be it. If there was ever a time when a new "trick" which would enhance survival would be strongly selected for this must be it. Recall Ridley's "...ecological ace ..." comments above.

In terms of survival, trade-mediated systems (liberalism-paradigm) are vastly more viable than are (realist paradigm) systems in which interactions between nations are mediated *primarily* by power contests. Systems which emphasize between-group cooperation unquestionably possess a clear advantage over those which rely on between-group competition, with cooperation reserved to within-group interactions. At this point I'd like to pose a related question: If this is so then why hasn't nature yet exploited it? Why don't, for example, ant colonies, trade? The answer to this question provides the answer to which model system variant most closely corresponds to the real world: All previous interaction systems, whether human, animal, or insect, have been *low power* systems. It is only at *very high levels of power* that the advantages of between-group cooperation become manifest. As, until very recently, the world system had been a (relatively) low power system, these advantages were not manifest. Thus, it closely approximated a pure realist paradigm, power-mediated system. Only recently have ever more rapid increases in systemic power begun to alter this prior reality. Recall that the metric of the system's learning space is endogenously generated. Power-driven systems must, therefore, possess quantifiably different metrics than do trade-driven systems. As discussed above, this is indeed the case. Both model variants seem, despite their differently operationalized assumptions, to be trying to tell a consistent story. The core elements of this story are that the world system is indeed, a complex adaptive learning system. In this context the system evinces substantially different attributes for high power configurations, than it does for low power ones. Much of the contemporary debate between realists, in all of their variants, and liberals/idealists, in all of their variants, can be attributed to a failure to comprehend the reality, and consequent implications of, this dynamic continuum of systemic adaptiveness.

CONCLUSION

In summation, the world system is a complex adaptive learning system. It is presently engaged in a rapid race between learning and collapse. Under conditions of temporally ever more swiftly increasing power growth it is experiencing increasingly strong 'selection pressure' to reconfigure itself in such a manner as to take advantage of the survival opportunities offered by between-group cooperation. This is the 'crisis' that my study has focused upon. It is worthwhile to recall that the Chinese ideogram for crisis is created by conjoining the otherwise separate characters for 'danger' and 'opportunity.' As such high-power systemically generated conditions have never previously existed, we are located in a unique moment in history—not only human history, but indeed the entire history of life on Earth. This is our contemporary danger—as well as our opportunity.

STUDY MODEL DIFFERENCE EQUATIONS

1) Relation: $R^{ij}_{(t+1)} = R^{ij}_{(t)} + b$;

$$\text{where } b = (.25 * ((RC^i * RC^j)/2)) * \frac{(R^{ij}_{(t)} - (R^{i+j}))}{(N-1)}$$

2) Reciprocity: $RC^i_{(t+1)} = 2 * \frac{(\sum {}^+R^i_{(t)})}{(\sum |R^i_{(t)}|)}$

3) Power: $P^{i'}_{(t+1)} = P^i_{(t)} + a$,

$$\text{where } a = \frac{|(\sum {}^+R^i_{(t)})|}{|(\sum R^i_{(t)})|} + \frac{|[Q * (\sum {}^+R_{(t)})]|}{|(\sum R_{(t)})|}$$

$$\text{where } Q = |P^i| + [P^{total}/(45 - (5 * RC^{total})].$$

4) Coupling: If $RC^i < MRC$: $CF^i = MRC - ((RC^i - MRC)/2)$

If $RC^i > MRC$: $CF^i = MRC + ((RC^i - MRC)/2)$

If $RC^i = MRC$: $CF^i = MRC$

where $MRC = RC^{total}/(N-1)$

Bibliography to Appendix

Allport, Allan, (1991) Visual Attention; in *Foundations of Cognitive Science*, Posner, Michael I, Ed, MIT Press, Cambridge MA.

Axelrod, R, (1984) *The Evolution of Cooperation*. Basic Books, New York, NY.

Baars, Bernard J, (1997) *In the Theater of Consciousness the Workshops of the Mind*. Oxford University Press, New York, NY.

Byron, Michael P, (1996) *Crisis-Driven Evolutionary Learning: Conceptual Foundations and Systemic Modeling*, Ph.D. Dissertation, UMI, Ann Arbor MI.

_____, (1997) Crisis-Driven Evolutionary Learning: A Procedural Mechanism for Systemic Learning, *Journal of Social and Evolutionary Systems*, 20(2): 179-83.

Casti, John I, (1996) *Would Be Worlds How Simulation is Changing the Frontiers of Science*. John Wiley & Sons Inc. New York, NY.

Chirot, Daniel, (1985) The Rise of the West, *American Sociological Review*, 50: 181-94.

Dennet, Daniel, (1995) *Darwin's Dangerous Idea Evolution and the Meaning of Life*, Simon and Schuster. New York, NY.

Doyle, Michael W, (1983) Kant, Liberal Legacies and Foreign Affairs, Part I, *Philosophy and Public Affairs*, 12(3): 205-35.

_____ (1986) Liberalism in World Politics, *American Political Science Review*, 80(4): 1151-69.

Eckstein, Harry, (1997) Culture as a Foundation Concept for the Social Sciences, *Journal of Theoretical Politics*, 8(40): 471-97.

Gilpin, Robert, (1981) *War and Change in World Politics*, Cambridge University Press, New York, NY.

Jervis, Robert, (1976) *Perception and Misperception in International Politics*, Princeton University Press, Princeton, NJ.

Levy, Jack, S, (1983) *War in the Modern Great Power System 1495-1975*. The University Press of Kentucky. Lexington, KY.

_____ (1994) Learning and Foreign Policy: Sweeping a Conceptual Minefield, *International Organization*, 48(20): 279-312.

Haas, Ernst, (1990) *When Knowledge is Power*. University of California Press, Berkley, CA.

Mansfield, Edward, D, (1994) *Power, Trade, and War*, Princeton University Press, Princeton, NJ.

Modelski, George, (1987) Is World Politics Evolutionary Learning? *International Organization*, 44(1): 1-23.

_____ (1996), Evolutionary Paradigm for Global Politics. *International Studies Quarterly*, 40: 321-342.

Ridley, Matthew, (1996) *The Origins of Virtue Human Instincts and the Evolution of Cooperation*. Penguin Books USA Inc, New York, NY.

Rumelhart, David E, (1991) The Architecture of the Mind: a Connectionist Approach; in *Foundations of Cognitive Science*, Posner, Michael I, Ed, MIT Press, Cambridge MA, 133-60.

Schater, Daniel, (1991) Memory; in *Foundations of Cognitive Science*, Posner, Michael I, Ed, MIT Press, Cambridge MA, 683-726.

Simon, Herbert A, (1985) Human Nature in Politics; The Dialogue of Psychology With Political Science, *American Political Science Review*, 82: 293-304.

Simon, Herbert A, & Kaplan, Craig A, (1991) Foundations of Cognitive Science; in *Foundations of Cognitive Science*, Posner, Michael I, Ed, MIT Press, Cambridge MA, 1-47.

Tetlock, Philip E, & Belkin, Aaron, (1996) *Counterfactual Thought Experiments in World Politics Logical, Methodological, and Psychological Perspectives*. Princeton University Press, Princeton NJ.

INDEX